Matrix Functions and Matrix Equations

Series in Contemporary Applied Mathematics   CAM

1   Jiongmin Yong, Rama Cont, eds., *Mathematical Finance: Theory and Practice* (2000)
2   F. Dubois, Huamo Wu, eds., *New Advances in Computational Fluid Dynamics: Theory, Methods and Applications* (2001)
3   Hanji Shang, Alain Tosseti, eds., *Actuarial Science: Theory and Practice* (2002)
4   Alexandre Ern, Weiping Liu, eds., *Mathematical Problems in Environmental Science and Engineering* (2002)
5   Haïm Brezis, Tatsien Li, eds., *Ginzburg-Landau Vortices* (2005)
6   Tatsien Li, Pingwen Zhang, eds., *Frontiers and Prospects of Contemporary Applied Mathematics* (2006)
7   Deguan Wang, Christian Duquennoi, Alexandre Ern, eds., *Mathematical Methods for Surface and Subsurface Hydrosystems* (2007)
8   Rolf Jeltsch, Tatsien Li, Ian H. Sloan, eds., *Some Topics in Industrial and Applied Mathematics* (2007)
9   Philippe G. Ciarlet, Tatsien Li, eds., *Differential Geometry: Theory and Applications* (2008)
10   Tatsien Li, Pingwen Zhang, eds., *Industrial and Applied Mathematics in China* (2009)
11   Zhien Ma, Yicang Zhou, Jianhong Wu, eds., *Modeling and Dynamics of Infectious Diseases* (2009)
12   Tomas Y. Hou, Chun Liu, Jianguo Liu, eds., *Multi-Scale Phenomena in Complex Fluids: Modeling, Analysis and Numerical Simulation* (2010)
13   Guiqiang Chen, Tatsien Li, Chun Liu, eds., *Nonlinear Conservation Laws, Fluid Systems and Related Topics* (2010)
14   A. Damlamian, S. Jaffard, eds., *Wavelet Methods in Mathematical Analysis and Engineering* (2010)
15   Tatsien Li, Yuejun Peng, Bopeng Rao, eds., *Some Problems on Nonlinear Hyperbolic Equations and Applications* (2010)
16   A. Damlamian, B. Miara, Tatsien Li, eds., *Multiscale Problems: Theory, Numerical Approximation and Applications* (2011)
17   Tatsien Li, Song Jiang, eds., *Hyperbolic Problems: Theory, Numerics and Applications*, Vol. 1 (2012)
18   Tatsien Li, Song Jiang, eds., *Hyperbolic Problems: Theory, Numerics and Applications*, Vol. 2 (2012)
19   Zhaojun Bai, Weiguo Gao, Yangfeng Su, *Matrix Functions and Matrix Equations* (2015)

Series in Contemporary Applied Mathematics CAM 19

Matrix Functions and Matrix Equations

editors

Zhaojun Bai
UC Davis

Weiguo Gao
Fudan University, China

Yangfeng Su
Fudan University, China

Higher Education Press

World Scientific

NEW JERSEY · LONDON · SINGAPORE · BEIJING · SHANGHAI · HONG KONG · TAIPEI · CHENNAI

Published by

Higher Education Press Limited Company
4 Dewai Dajie, Beijing 100120, P. R. China
and
World Scientific Publishing Co. Pte. Ltd.
5 Toh Tuck Link, Singapore 596224

Library of Congress Cataloging-in-Publication Data
Matrix functions and matrix equations / edited by Zhaojun Bai (UC Davis), Weiguo Gao (Fudan
University, China), Yangfeng Su (Fudan University, China).
    148 pages cm. -- (Series in contemporary applied mathematics ; vol. 19)
    Includes bibliographical references.
    ISBN 978-9814675765 (hardcover : alk. paper)
    1. Matrices. 2. Linear operators. 3. Algebras, Linear. I. Bai, Zhaojun, editor. II. Gao, Weiguo,
editor. III. Su, Yangfeng, editor.
    QA184.2.M39 2015
    512.9'434--dc23
                                    2015005054

British Library Cataloguing-in-Publication Data
A catalogue record for this book is available from the British Library.

Preface

This volume is a collection of summaries from the lectures of the Fourth Gene Golub SIAM Summer School on "Matrix Functions and Matrix Equations" held at Fudan University, Shanghai, China from July 22 to August 2, 2013. The School was in conjunction with the 3rd International Summer School on Numerical Linear Algebra and the 9th Shanghai Summer School on Analysis and Numerics in Modern Sciences. 45 students from 14 countries attended the School. An extra week of activities from August 5 to August 9 was organized for interested students.

Matrix functions and matrix equations are widely used in science, engineering and the social sciences, due to the succinct and insightful way in which they allow problems to be formulated and solutions to be expressed. Applications range from exponential integrators for the solution of partial differential equations to model reduction of dynamical systems. The School introduces students to underlying theory, algorithms and applications of matrix functions and matrix equations, and relevant linear solvers and eigenvalue computations. The summer school was composed of three courses:

1. Functions of matrices and exponential integrators by Nicholas Higham, The University of Manchester, United Kingdom and Marlis Hochbruck, Karlsruhe Institute of Technology, Germany.
2. Matrix equations and model reduction by Peter Benner, Max-Planck Institute for Dynamics of Complex Technical Systems, Magdeburg, Germany.
3. High performance linear solvers and eigenvalue computations by Ren-Cang Li, University of Texas at Arlington, United States and Xiaoye Sherry Li, Lawrence Berkeley National Laboratory, United States.

There are five chapters in this volume. Chapter 1, by Nicholas Higham and Lijing Lin, is on matrix functions: a short course. Chapter 2, by Marlis Hochbruck, is on a short course on exponential integrators. Chapter 3, by Peter Benner, Tobias Breiten and Lihong Feng, is on matrix equations and model reduction. Chapter 4, by Ren-Cang Li, is on Rayleigh quotient based optimization methods for eigenvalue problems. Chapter 5, by Xiaoye Li, is on factorization based sparse solvers and preconditioners.

The summer school was hosted by School of Mathematical Sciences, Fudan University. The local organizers include Professors Tatsien Li, Jin Cheng, Weiguo Gao and Yangfeng Su of Fudan University, Professor Pingwen Zhang of Peking University and Professor Zhaojun Bai of University of California, Davis.

The School received the generous sponsorship from SIAM, US National Science Foundation, Chinese-French Institute for Applied Mathematics (ISFMA), Shanghai Center for Mathematical Sciences, National Science Foundation of China (NSFC), and NSFC 111 project. In addition, many units of Fudan University including Graduate School, DOE Key Laboratory of Nonlinear Mathematical Models and Methods, Key Laboratory of Shanghai Modern Applied Mathematics also provided financial and logistical supports. The Numerical Algorithm Group (NAG) provided the free access to numerical computing software for all participants.

We would like to express our gratitude to all authors for their contributions to this volume. Finally, the editors wish to thank Mr. Tianfu Zhao for the professional assistant on the publication of this volume.

October 2014

Zhaojun Bai
Weiguo Gao
Yangfeng Su

Contents

Matrix Functions: A Short Course

Nicholas J. Higham*   Lijing Lin[†]

1 Introduction

A summary is given of a course on functions of matrices delivered by the first author (lecturer) and second author (teaching assistant) at the Gene Golub SIAM Summer School 2013 at Fudan University, Shanghai, China, July 22–26 2013 [34]. This article covers some essential features of the theory and computation of matrix functions. In the spirit of course notes the article is not a comprehensive survey and does not cite all the relevant literature. General references for further information are the book on matrix functions by Higham [31] and the survey by Higham and Al-Mohy [35] of computational methods for matrix functions.

2 History

Matrix functions are as old as matrix algebra itself. The term "matrix" was coined in 1850 [58] by James Joseph Sylvester, FRS (1814–1897), while the study of matrix algebra was initiated by Arthur Cayley, FRS (1821–1895) in his *A Memoir on the Theory of Matrices* (1858) [11]. In that first paper, Cayley considered matrix square roots.

Notable landmarks in the history of matrix functions include:

- Laguerre (1867) [45] and Peano (1888) [53] defined the exponential of a matrix via its power series.

- Sylvester (1883) stated the definition of $f(A)$ for general f via the interpolating polynomial [59]. Buchheim (1886) [10], [33] extended Sylvester's interpolation formula to arbitrary eigenvalues.

- Frobenius (1896) [21] showed that if f is analytic then $f(A)$ is the sum of the residues of $(zI - A)^{-1}f(z)$ at the eigenvalues of A, thereby anticipating the Cauchy integral representation, which was used by Poincaré (1899) [54].

School of Mathematics, The University of Manchester, Manchester, M13 9PL, UK (*nick.higham@manchester.ac.uk, http://www.maths.man.ac.uk/~higham; [†]lijing.lin@manchester.ac.uk, http://www.maths.manchester.ac.uk/~lijing)

- The Jordan form definition was used by Giorgi (1928) [22], and Cipolla (1932) [14] extended it to produce nonprimary matrix functions.

- The first book on matrix functions was published by Schwerdtfeger (1938) [56].

- Frazer, Duncan and Collar published the book *Elementary Matrices and Some Applications to Dynamics and Differential Equations* [20] in 1938, which was "the first book to treat matrices as a branch of applied mathematics" [15].

- A research monograph on functions of matrices was published by Higham (2008) [31].

3 Theory

3.1 Definitions

We are concerned with functions $f : \mathbb{C}^{n \times n} \to \mathbb{C}^{n \times n}$ that are defined in terms of an underlying scalar function f. Given $f(t)$, one can define $f(A)$ by substituting A for t: e.g.,

$$f(t) = \frac{1 + t^2}{1 - t} \quad \Rightarrow \quad f(A) = (I - A)^{-1}(I + A^2),$$

$$\log(1 + x) = x - \frac{x^2}{2} + \frac{x^3}{3} - \frac{x^4}{4} + \cdots, \quad |x| < 1,$$

$$\Rightarrow \log(I + A) = A - \frac{A^2}{2} + \frac{A^3}{3} - \frac{A^4}{4} + \cdots, \quad \rho(A) < 1.$$

This way of defining $f(A)$ works for f a polynomial, a rational function, or a function having a convergent power series (see section 6.1). Note that f is not evaluated elementwise on the matrix A, as is the case in some programming languages.

For general f, there are various equivalent ways to formally define a matrix function. We give three definitions, based on the Jordan canonical form, polynomial interpolation, and the Cauchy integral formula.

3.1.1 Definition via Jordan canonical form

Any matrix $A \in \mathbb{C}^{n \times n}$ can be expressed in the Jordan canonical form

$$Z^{-1}AZ = J = \text{diag}(J_1, J_2, \ldots, J_p), \tag{3.1a}$$

$$J_k = J_k(\lambda_k) = \begin{bmatrix} \lambda_k & 1 & & \\ & \lambda_k & \ddots & \\ & & \ddots & 1 \\ & & & \lambda_k \end{bmatrix} \in \mathbb{C}^{m_k \times m_k}, \tag{3.1b}$$

where Z is nonsingular and $m_1 + m_2 + \cdots + m_p = n$. Denote by

- $\lambda_1, \ldots, \lambda_s$ the distinct eigenvalues of A,
- n_i the order of the largest Jordan block in which λ_i appears, which is called the *index* of λ_i.

We say the function f is *defined on the spectrum* of A if the values

$$f^{(j)}(\lambda_i), \qquad j = 0 \colon n_i - 1, \quad i = 1 \colon s \tag{3.2}$$

exist.

Definition 3.1 (matrix function via Jordan canonical form). *Let f be defined on the spectrum of $A \in \mathbb{C}^{n \times n}$ and let A have the Jordan canonical form* (3.1). *Then*

$$f(A) := Z f(J) Z^{-1} = Z \operatorname{diag}(f(J_k)) Z^{-1}, \tag{3.3}$$

where

$$f(J_k) := \begin{bmatrix} f(\lambda_k) & f'(\lambda_k) & \cdots & \dfrac{f^{(m_k-1)}(\lambda_k)}{(m_k - 1)!} \\ & f(\lambda_k) & \ddots & \vdots \\ & & \ddots & f'(\lambda_k) \\ & & & f(\lambda_k) \end{bmatrix}. \tag{3.4}$$

The definition yields a matrix $f(A)$ that can be shown to be independent of the particular Jordan canonical form.

In the case of multivalued functions such as $\sqrt{t}$ and $\log t$ it is implicit that a single branch has been chosen in (3.4) and across the Jordan blocks with the same eigenvalue; the resulting function is called a *primary* matrix function. If an eigenvalue occurs in more than one Jordan block and a different choice of branch is made in two different blocks, then a *nonprimary* matrix function is obtained (see section 3.1.5).

3.1.2 Definition via interpolating polynomial

Before giving the second definition, we recall some background on polynomials at a matrix argument.

- The *minimal polynomial* of $A \in \mathbb{C}^{n \times n}$ is defined to be the unique monic polynomial ϕ of lowest degree such that $\phi(A) = 0$. The existence of the minimal polynomial is proved in most textbooks on linear algebra.
- By considering the Jordan canonical form, it is not hard to see that $\phi(t) = \prod_{i=1}^{s} (t - \lambda_i)^{n_i}$, where $\lambda_1, \ldots, \lambda_s$ are the distinct eigenvalues of A and n_i is the index of λ_i. It follows immediately that ϕ is zero on the spectrum of A (that is, the values (3.2) are all zero for $f(t) = \phi(t)$).

- Given any polynomial p and any matrix $A \in \mathbb{C}^{n \times n}$, p is clearly defined on the spectrum of A and $p(A)$ can be defined by substitution.

- For polynomials p and q, $p(A) = q(A)$ if and only if p and q take the same values on the spectrum [31, Thm. 1.3]. Thus the matrix $p(A)$ is completely determined by the values of p on the spectrum of A.

The following definition gives a way to generalize the property of polynomials in the last bullet point to arbitrary functions and define $f(A)$ in terms of the values of f on the spectrum of A.

Definition 3.2 (matrix function via Hermite interpolation). *Let f be defined on the spectrum of $A \in \mathbb{C}^{n \times n}$. Then $f(A) := p(A)$, where p is the unique polynomial of degree less than $\sum_{i=1}^{s} n_i$ (which is the degree of the minimal polynomial) that satisfies the interpolation conditions*

$$p^{(j)}(\lambda_i) = f^{(j)}(\lambda_i), \qquad j = 0 \colon n_i - 1, \quad i = 1 \colon s.$$

The polynomial p specified in the definition is known as the Hermite interpolating polynomial.

For an example, let $f(t) = t^{1/2}$ (the principal branch of the square root function, so that $\operatorname{Re} t^{1/2} \geq 0$), $A = \left[\begin{smallmatrix} 2 & 2 \\ 1 & 3 \end{smallmatrix}\right]$, $\lambda(A) = \{1, 4\}$. Seeking $p(t)$ with $p(1) = f(1)$ and $p(4) = f(4)$, we obtain

$$p(t) = f(1)\frac{t-4}{1-4} + f(4)\frac{t-1}{4-1} = \frac{1}{3}(t+2)$$

$$\Rightarrow \quad A^{1/2} = p(A) = \frac{1}{3}(A + 2I) = \frac{1}{3}\begin{bmatrix} 4 & 2 \\ 1 & 5 \end{bmatrix}.$$

Several properties follow immediately from this definition:

- $f(A) = p(A)$ is a polynomial in A, where the polynomial p depends on A.

- $f(A)$ commutes with A.

- $f(A^T) = f(A)^T$.

Because the minimal polynomial divides the characteristic polynomial, $q(t) = \det(tI - A)$, it follows that $q(A) = 0$, which is the Cayley-Hamilton theorem. Hence A^n can be expressed as a linear combination of lower powers of A: $A^n = \sum_{k=0}^{n-1} c_k A^k$. Using this relation recursively we find that any power series collapses to a polynomial. For example, $e^A = \sum_{k=0}^{\infty} A^k/k! = \sum_{k=0}^{n-1} d_k A^k$ (but the d_k depend on A).

3.1.3 Definition via Cauchy integral theorem

Definition 3.3 (matrix function via Cauchy integral). *For $A \in \mathbb{C}^{n \times n}$*,

$$f(A) = \frac{1}{2\pi i} \int_\Gamma f(z)(zI - A)^{-1} dz,$$

where f is analytic on and inside a closed contour Γ that encloses $\lambda(A)$.

3.1.4 Multiplicity and equivalence of definitions

Definitions 3.1, 3.2, and 3.3 are equivalent, modulo the analyticity assumption for the Cauchy integral definition [31, Thm. 1.12]. Indeed this equivalence extends to other definitions, as noted by Rinehart [55]:

"There have been proposed in the literature since 1880 eight distinct definitions of a matric function, by Weyr, Sylvester and Buchheim, Giorgi, Cartan, Fantappiè, Cipolla, Schwerdtfeger and Richter ... All of the definitions, except those of Weyr and Cipolla are essentially equivalent."

The definitions have different strengths. For example, the interpolation definition readily yields some key basic properties (as we have already seen), the Jordan canonical form definition is useful for solving matrix equations (e.g., $X^2 = A$, $e^X = A$) and for evaluation when A is normal, and the Cauchy integral definition can be useful both in theory and in computation (see section 7.2).

3.1.5 Nonprimary matrix functions

Nonprimary matrix functions are those that are not obtainable from our three definitions, or that violate the single branch requirement in Definition 3.1. Thus a nonprimary matrix function of A is obtained from Definition 3.1 if A is derogatory and a different branch of f is taken in two different Jordan blocks for λ. For example, the 2×2 identity matrix has two primary square roots and an infinity of nonprimary square roots:

$$I_2 = \begin{bmatrix} 1 & 0 \\ 0 & 1 \end{bmatrix}^2 = \begin{bmatrix} -1 & 0 \\ 0 & -1 \end{bmatrix}^2 \quad \text{(primary)}$$

$$= \begin{bmatrix} 1 & 0 \\ 0 & -1 \end{bmatrix}^2 = \begin{bmatrix} \cos\theta & \sin\theta \\ \sin\theta & -\cos\theta \end{bmatrix}^2 \quad \text{(nonprimary)}.$$

In general, primary matrix functions are expressible as polynomials in A, while nonprimary ones are not. The 2×2 zero matrix 0_2 is its own primary square root. Any nilpotent matrix of degree 2 is also a nonprimary square root of 0_2, for example $\left[\begin{smallmatrix} 0 & 1 \\ 0 & 0 \end{smallmatrix}\right]$, but the latter matrix is not a polynomial in 0_2.

The theory of matrix functions is almost exclusively concerned with primary matrix functions, but nonprimary functions are needed in some applications, such as the embeddability problem in Markov chains [31, Sec. 2.3].

3.1.6 Principal logarithm, root, and power

Let $A \in \mathbb{C}^{n \times n}$ have no eigenvalues on $\mathbb{R}^-$ (the closed real axis). We need the following definitions.

Principal log: $X = \log A$ denotes the unique X such that $e^X = A$ and $-\pi < \operatorname{Im} \lambda_i < \pi$ for every eigenvalue λ_i of X.

Principal pth root: For integer $p > 0$, $X = A^{1/p}$ is the unique X such that $X^p = A$ and $-\pi/p < \arg \lambda_i < \pi/p$ for every eigenvalue λ_i of X.

Principal power: For $s \in \mathbb{R}$, the principal power is defined as $A^s = e^{s \log A}$, where $\log A$ is the principal logarithm. An integral representation is also available:

$$A^s = \frac{\sin(s\pi)}{s\pi} A \int_0^\infty (t^{1/s} I + A)^{-1} dt, \qquad s \in (0,1).$$

3.2   Properties and formulas

Three basic properties of $f(A)$ were stated in section 3.1.2. Some other important properties are collected in the following theorem.

Theorem 3.4 ([31, Thm. 1.13]). *Let $A \in \mathbb{C}^{n \times n}$ and let f be defined on the spectrum of A. Then*

(a) $f(XAX^{-1}) = Xf(A)X^{-1}$;

(b) *the eigenvalues of $f(A)$ are $f(\lambda_i)$, where the λ_i are the eigenvalues of A;*

(c) *if X commutes with A then X commutes with $f(A)$;*

(d) *if $A = (A_{ij})$ is block triangular then $F = f(A)$ is block triangular with the same block structure as A, and $F_{ii} = f(A_{ii})$;*

(e) *if $A = \operatorname{diag}(A_{11}, A_{22}, \dots, A_{mm})$ is block diagonal then $f(A) = \operatorname{diag}(f(A_{11}), f(A_{22}), \dots, f(A_{mm}))$.*

Some more advanced properties are as follows.

- $f(A) = 0$ if and only if (from Definition 3.1 or 3.2) $f^{(j)}(\lambda_i) = 0$, $j = 0 : n_i - 1$, $i = 1 : s$.

- The sum, product, composition of functions work "as expected":
  - $(\sin + \cos)(A) = \sin A + \cos A$,
  - $f(t) = \cos(\sin t) \quad \Rightarrow \quad f(A) = \cos(\sin A)$.
- Polynomial functional relations generalize from the scalar case. For example: if $G(f_1, \ldots, f_m) = 0$, where G is a polynomial, then $G(f_1(A), \ldots, f_m(A)) = 0$. E.g.,
  - $\sin^2 A + \cos^2 A = I$,
  - $(A^{1/p})^p = A$ for any integer $p > 0$,
  - $e^{iA} = \cos A + i \sin A$.
- However, other plausible relations can fail:
  - $f(A^*) \neq f(A)^*$ in general,
  - $e^{\log A} = A$ but $\log e^A \neq A$ in general,
  - $e^A \neq (e^{A/\alpha})^\alpha$ in general,
  - $(AB)^{1/2} \neq A^{1/2}B^{1/2}$ in general,
  - $e^{(A+B)t} = e^{At}e^{Bt}$ for all t if and only if $AB = BA$.

Correction terms involving the matrix unwinding function can be introduced to restore equality in the second to fourth cases [7].

3.3 Fréchet derivative and condition number

3.3.1 Relative condition number

An important issue in the computation of matrix functions is the conditioning. The data may be uncertain and rounding errors from finite precision computations can often be interpreted via backward error analysis as being equivalent to perturbations in the data. So it is important to understand the sensitivity of $f(A)$ to perturbations in A. Sensitivity is measured by the condition number defined as follows.

Definition 3.5. *Let* $f : \mathbb{C}^{n \times n} \to \mathbb{C}^{n \times n}$ *be a matrix function. The relative condition number of* f *is*

$$\mathrm{cond}(f, A) := \lim_{\epsilon \to 0} \sup_{\|E\| \leq \epsilon \|A\|} \frac{\|f(A + E) - f(A)\|}{\epsilon \|f(A)\|},$$

where the norm is any matrix norm.

3.3.2 Fréchet derivative

To obtain explicit expressions for $\mathrm{cond}(f, A)$, we need an appropriate notion of derivative for matrix functions. The *Fréchet derivative* of a matrix function $f : \mathbb{C}^{n \times n} \to \mathbb{C}^{n \times n}$ at a point $A \in \mathbb{C}^{n \times n}$ is a linear mapping $L_f(A, \cdot) : \mathbb{C}^{n \times n} \to \mathbb{C}^{n \times n}$ such that for all $E \in \mathbb{C}^{n \times n}$

$$f(A + E) = f(A) + L_f(A, E) + o(\|E\|).$$

It is easy to show that the condition number $\mathrm{cond}(f, A)$ can be characterized as

$$\mathrm{cond}(f, A) = \frac{\|L_f(A)\|\,\|A\|}{\|f(A)\|},$$

where

$$\|L_f(A)\| := \max_{Z \neq 0} \frac{\|L_f(A, Z)\|}{\|Z\|}.$$

3.3.3 Condition number estimation

Since L_f is a linear operator,

$$\mathrm{vec}(L_f(A, E)) = K_f(A)\mathrm{vec}(E)$$

where $K_f(A) \in \mathbb{C}^{n^2 \times n^2}$ is a matrix independent of E known as the *Kronecker form* of the Fréchet derivative. It can be shown that $\|L_f(A)\|_F = \|K_f(A)\|_2$ and that $\|L_f(A)\|_1$ and $\|K_f(A)\|_1$ differ by at most a factor n. Hence estimating $\mathrm{cond}(f, A)$ reduces to estimating $\|K_f(A)\|$ and this can be done using a matrix norm estimator, such as the block 1-norm estimator of Higham and Tisseur [39].

4 Applications

Functions of matrices play an important role in many applications. Here we describe some examples.

4.1 Toolbox of matrix functions

In software we want to be able to evaluate interesting f at matrix arguments as well as scalar arguments. For example, trigonometric matrix functions, as well as matrix roots, arise in the solution of second order differential equations: the initial value problem

$$\frac{d^2 y}{dt^2} + Ay = 0, \quad y(0) = y_0, \quad y'(0) = y_0'$$

has solution

$$y(t) = \cos(\sqrt{A}t)y_0 + (\sqrt{A})^{-1}\sin(\sqrt{A}t)y_0',$$

where $\sqrt{A}$ denotes any square root of A. On the other hand, the differential equation can be converted to a first order system and then solved using the exponential:

$$\begin{bmatrix} y' \\ y \end{bmatrix} = \exp\left(\begin{bmatrix} 0 & -tA \\ tI_n & 0 \end{bmatrix}\right) \begin{bmatrix} y_0' \\ y_0 \end{bmatrix}.$$

4.2 Nuclear magnetic resonance

In nuclear magnetic resonance (NMR) spectroscopy, the Solomon equations

$$\frac{dM}{dt} = -RM, \quad M(0) = I$$

relate a matrix of intensities $M(t)$ to a symmetric, diagonally dominant matrix R (known as the relaxation matrix). Hence $M(t) = e^{-Rt}$. NMR workers need to solve both forward and inverse problems:

- in simulations and testing, compute $M(t)$ given R;

- determine R from observed intensities: estimation methods are used since not all the m_{ij} are observed.

4.3 Phi functions and exponential integrators

The φ functions are defined by the recurrence $\varphi_{k+1}(z) = \frac{\varphi_k(z) - 1/k!}{z}$ with $\varphi_0(z) = e^z$, and are given explicitly by

$$\varphi_k(z) = \sum_{j=0}^{\infty} \frac{z^j}{(j+k)!}.$$

They appear in explicit solutions to certain linear differential equations:

$$\frac{dy}{dt} = Ay, \quad y(0) = y_0 \quad \Rightarrow \quad y(t) = e^{At} y_0,$$

$$\frac{dy}{dt} = Ay + b, \quad y(0) = 0 \quad \Rightarrow \quad y(t) = t\varphi_1(tA)b,$$

$$\frac{dy}{dt} = Ay + ct, \quad y(0) = 0 \quad \Rightarrow \quad y(t) = t^2\varphi_2(tA)c,$$

and more generally provide an explicit solution for a differential equation with right-hand side $Ay + p(t)$ with p a polynomial.

Consider an initial value problem written in the form

$$u'(t) = Au(t) + g(t, u(t)), \quad u(t_0) = u_0, \quad t \geq t_0, \tag{4.1}$$

where $u(t) \in \mathbb{C}^n$, $A \in \mathbb{C}^{n \times n}$, and g is a nonlinear function. Spatial semidiscretization of partial differential equations leads to systems in this form with A representing a discretized linear operator. Thus A may be large and sparse. The solution can be written as [47, Lem. 5.1]

$$u(t) = e^{(t-t_0)A} u_0 + \sum_{k=1}^{\infty} \varphi_k\big((t-t_0)A\big)(t-t_0)^k u_k, \tag{4.2}$$

where

$$u_k = \frac{d^{k-1}}{dt^{k-1}} g(t, u(t)) \mid_{t=t_0}.$$

By truncating the series in (4.2), we obtain the approximation

$$u(t) \approx \widehat{u}(t) = e^{(t-t_0)A} u_0 + \sum_{k=1}^{p} \varphi_k \big((t-t_0)A\big)(t-t_0)^k u_k. \qquad (4.3)$$

Exponential integrator methods are obtained by employing suitable approximations to the vectors u_k [40]. The simplest method is the exponential time differencing (ETD) Euler method [46]

$$y_n = e^{hA} y_{n-1} + h\varphi_1(hA) g(t_{n-1}, y_{n-1}).$$

Clearly, implementing an exponential integrator involves computing matrix–vector products involving the exponential and the φ functions.

4.4   Complex networks

Let A be an adjacency matrix of an undirected network. Certain characteristics of the network are defined in terms of the matrix exponential [19]

$$e^A = I + A + \frac{1}{2}A^2 + \frac{1}{3!}A^3 + \frac{1}{4!}A^4 + \cdots.$$

The *centrality* of node i, defined as $(e^A)_{ii}$, measures how important that node is. The resolvent $(I - \alpha A)^{-1}$ can be used in place of e^A, as was originally done by Katz [41]. The *communicability* between nodes i and j, defined as $(e^A)_{ij}$, measures how well information is transferred between the nodes.

4.5   Random multivariate samples in statistics

Suppose we wish to generate random vectors distributed as $y \sim N(\mu, C)$, where $N(\mu, C)$ denotes the multivariate normal distribution with mean μ and covariance matrix C. This can be done by generating standard normally distributed vectors $x \sim N(0, I)$ and setting $y = \mu + Lx$, where $C = LL^T$ is the Cholesky factorization. However, in some applications C has dimension greater than 10^{12} and computing the Cholesky factor is impractical. Chen, Anitescu, and Saad [12] note that one may instead generate y as $y = \mu + C^{1/2}x$. Methods are available for computing $C^{1/2}x$ that require only matrix–vector products with C (see sections 7.2 and 7.3) and so this computation is feasible, even if computing the Cholesky factor is not.

4.6 The average eye in optics

The first order character of an optical system is characterized by the *transference matrix*

$$T = \begin{bmatrix} S & \delta \\ 0 & 1 \end{bmatrix} \in \mathbb{R}^{5 \times 5},$$

where $S \in \mathbb{R}^{4 \times 4}$ is symplectic, that is,

$$S^T J S = J = \begin{bmatrix} 0 & I_2 \\ -I_2 & 0 \end{bmatrix}.$$

A straightforward average $m^{-1} \sum_{i=1}^{m} T_i$ is not in general a transference matrix. Harris [26] proposes as a suitable average the matrix $\exp\left(m^{-1} \sum_{i=1}^{m} \log T_i\right)$, where $\log$ is the principal logarithm, which is always a transference matrix.

5 Problem classification

The choice of method to compute $f(A)$ should take into account the properties of A, the size of the problem, the type of function, and accuracy requirements. We classify the problems according to their size and then describe some of the main methods for each class in sections 6 and 7.

5.1 Small/medium scale $f(A)$ problems

For this class of problems it is possible to compute a decomposition of A and to store $f(A)$.

For a normal matrix A we can compute the Schur (spectral) decomposition $A = QDQ^*$, with Q unitary and $D = \mathrm{diag}(d_i)$, and then form $f(A) = Q\mathrm{diag}(f(d_i))Q^*$. If A is nonnormal we can compute a Schur decomposition $A = QTQ^*$, with Q unitary and T upper triangular, and then the Schur–Parlett algorithm described in section 6.7 can be used.

For a number of matrix functions, such as matrix roots, the matrix sign function, and the unitary polar factor, $f(A)$ can be computed by a matrix iteration $X_{k+1} = g(X_k)$, $X_0 = A$, where g is some nonlinear function. Usually, g is rational and so the iterations require only matrix multiplication and the solution of multiple right-hand side linear systems.

Another important tool for evaluating matrix functions is approximation, whereby $f(A)$ is approximated by $r(A)$, where $r(x)$ is a polynomial or rational approximation to $f(x)$, such as a truncated Taylor series or a Padé approximant. In this case some preprocessing is needed to get A into a region where the approximant is sufficiently accurate.

5.2   Large scale $f(A)b$ problems

If A is sufficiently large and sparse then it will be undesirable or impossible to compute a Schur decomposition of A or store $f(A)$. Therefore the problem of interest is to compute $f(A)b$, the action of $f(A)$ on b, without first computing $f(A)$. There are in general two different cases.

- Case 1: we are able to solve $Ax = b$ by a sparse direct method. Then methods based on the Cauchy integral formula can be used. Also, rational Krylov methods can be used with direct solves.

- Case 2: we can only compute matrix-vector products Ax (and perhaps A^*x). Exponential integrators for sufficiently large problems are contained in this case. Krylov methods based on the Arnoldi or Lanczos process, or methods employing polynomial approximations, can be used.

5.3   Accuracy requirements

The desired accuracy is an important question when choosing a method. This may range from full double precision accuracy (about 16 significant decimal digits) to just 3 or 4 digits if the matrix A is subject to large measurement errors, as may be the case in problems in engineering or healthcare.

Some methods will accept the error tolerance as a parameter, while others are always designed to aim for full precision. Some methods work best when less than full precision is required.

A further consideration is that for testing purposes we may want to compute a very accurate solution that can be taken as the "exact solution". Thus we may need a method that can deliver a computed solution correct to quadruple precision, or even higher precision, when implemented in high precision arithmetic.

6   Methods for $f(A)$

6.1   Taylor series

The Taylor series is a basic tool for approximating matrix functions. If f has a Taylor series expansion

$$f(z) = \sum_{k=0}^{\infty} a_k (z - \alpha)^k$$

with radius of convergence r then for $A \in \mathbb{C}^{n \times n}$ the series

$$f(A) = \sum_{k=0}^{\infty} a_k (A - \alpha I)^k \tag{6.1}$$

converges if $|\lambda_i - \alpha| < r$ for every eigenvalue λ_i of A [31, Thm. 4.7]. The error in a truncated Taylor series with terms up to $(A - \alpha I)^{m-1}$ in (6.1) can be bounded in terms of the mth derivative of f at matrix arguments. Just as for scalar Taylor series, numerical cancellation must be avoided by restricting the size of $\|A - \alpha I\|$.

6.2 Padé approximation

For a given scalar function $f(x)$, the rational function

$$r_{km}(x) = p_{km}(x)/q_{km}(x)$$

is a $[k/m]$ *Padé approximant* of f if r_{km} has numerator and denominator polynomials of degrees at most k and m, respectively, $q_{km}(0) = 1$, and

$$f(x) - r_{km}(x) = O(x^{k+m+1}).$$

If a $[k/m]$ Padé approximant exists then it is unique. Padé approximants tend to be more efficient for matrix arguments than truncated Taylor series in that they can deliver similar accuracy at lower cost. For some important functions, such as the exponential and the logarithm, Padé approximants are explicitly known.

The choice of method to evaluate a Padé approximant at a matrix argument is based on balancing numerical stability and computational cost. Possibilities are

- a ratio of polynomials: $r_{km}(A) = q_{km}(A)^{-1} p_{km}(A)$, with an appropriate way to evaluate $p_{km}(A)$ and $q_{km}(A)$ (for example, Horner's method or the Paterson–Stockmeyer method [31, Sec. 4.2], [52]),

- continued fraction form, with bottom-up or top-down evaluation,

- partial fraction form.

6.3 Similarity transformations

Given a factorization $A = XBX^{-1}$, we can use the formula $f(A) = Xf(B)X^{-1}$, provided that $f(B)$ is easily computable, for example if $B = \operatorname{diag}(\lambda_i)$. However, any error ΔB in $f(B)$ can be magnified by as much as $\kappa(X) = \|X\|\|X^{-1}\| \geq 1$ in $f(A)$. Therefore we prefer to work with unitary X. Hence we typically use an eigendecomposition (diagonal B) when A is normal ($AA^* = A^*A$), or a Schur decomposition (triangular B) in general.

We could also take $B = \mathrm{diag}(B_i)$ block diagonal and require X to be well conditioned. Such a decomposition can be computed by starting with a Schur decomposition and then eliminating off-diagonal blocks using similarity transformations obtained by solving Sylvester equations. This approach needs a parameter: the maximum allowed condition number of individual transformations. The larger that parameter, the more numerous and smaller the diagonal blocks will be. The diagonal blocks B_i are triangular but have no particular eigenvalue distribution, so computing $f(B_i)$ is nontrivial. Block diagonalization has not proved to be a popular approach.

6.4 Schur method for matrix roots

Björck and Hammarling [9] show that a square root X of a matrix $A \in \mathbb{C}^{n \times n}$ can be computed by computing a Schur decomposition $A = QTQ^*$, solving $U^2 = T$ for the upper triangular matrix U by the recurrence

$$u_{ii} = \sqrt{t_{ii}}, \quad u_{ij} = \frac{t_{ij} - \sum_{k=i+1}^{j-1} u_{ik} u_{kj}}{u_{ii} + u_{jj}}, \tag{6.2}$$

and then forming $X = QUQ^*$. This method has essentially optimal numerical stability. It was extended to use the real Schur decomposition for real matrices by Higham [28] and to compute pth roots by Smith [57]. The recurrences for $p > 2$ are substantially more complicated than those for $p = 2$.

Recently, Deadman, Higham, and Ralha [17] have developed blocked versions of the recurrence (6.2) that give substantially better performance on modern computers.

6.5 Parlett's recurrence

If T is upper triangular then $F = f(T)$ is upper triangular and the diagonal elements of F are $f_{ii} = f(t_{ii})$. Parlett [51] shows that the off-diagonal elements of F can be obtained from the recurrence, derived from $FT = TF$,

$$f_{ij} = t_{ij} \frac{f_{ii} - f_{jj}}{t_{ii} - t_{jj}} + \sum_{k=i+1}^{j-1} \frac{f_{ik} t_{kj} - t_{ik} f_{kj}}{t_{ii} - t_{jj}},$$

which enables F to be computed a column or a superdiagonal at a time. The recurrence fails when T has repeated eigenvalues and can suffer severe loss of accuracy in floating point arithmetic when two eigenvalues

t_{ii} and t_{jj} are very close. A way around these problems is to employ a block form of this recurrence.

6.6 Block Parlett recurrence

For upper triangular T, Parlett [50] partitions $T = (T_{ij})$ with square diagonal blocks. Then F has the same block upper triangular structure and $F_{ii} = f(T_{ii})$. The equation $FT = TF$ leads to Sylvester equations

$$T_{ii}F_{ij} - F_{ij}T_{jj} = F_{ii}T_{ij} - T_{ij}F_{jj} + \sum_{k=i+1}^{j-1} (F_{ik}T_{kj} - T_{ik}F_{kj}),$$

which are nonsingular as long as no two different diagonal blocks T_{ii} and T_{jj} have an eigenvalue in common. Thus F can be computed a block superdiagonal or a block column at a time. We can expect numerical difficulties when two blocks have close spectra.

6.7 Schur–Parlett algorithm

Davies and Higham [16] build from the block Parlett recurrence a general purpose algorithm for computing $f(A)$. The key ideas are to reorder and re-block in an attempt to produce well conditioned Sylvester equations and to evaluate $f(T_{ii})$ from a Taylor series (unless a more specific method is available for the given f).

The outline of the Schur–Parlett algorithm is:

- Compute a Schur decomposition $A = QTQ^*$.

- Reorder T to block triangular form in which eigenvalues within a diagonal block are "close" and those of different diagonal blocks are "well separated".

- Evaluate $F_{ii} = f(T_{ii})$ for each i.

- Solve in an appropriate order the Sylvester equations

$$T_{ii}F_{ij} - F_{ij}T_{jj} = F_{ii}T_{ij} - T_{ij}F_{jj} + \sum_{k=i+1}^{j-1} (F_{ik}T_{kj} - T_{ik}F_{kj}).$$

- Undo the unitary transformations.

Reordering step. The eigenvalues are split into sets such that λ_i and λ_j go in the same set if, for some parameter $\delta > 0$, $|\lambda_i - \lambda_j| \leq \delta$. An ordering of sets on the diagonal is chosen and a sequence of swaps of diagonal elements to produce that ordering determined. The swaps are effected by unitary transformations [8].

Function of atomic diagonal block. Let $U \in \mathbb{C}^{m \times m}$ represent an atomic diagonal block of the reordered Schur form. Assume f has a Taylor series with an infinite radius of convergence and that all the derivatives are available. We write $U = \sigma I + M$, where $\sigma = \text{trace}(U)/m$ is the mean of the eigenvalues, and then take the Taylor series about σ:

$$f(U) = \sum_{k=0}^{\infty} \frac{f^{(k)}(\sigma)}{k!} M^k.$$

Because the convergence of the series can be very nonmonotonic we truncate it based on a strict error bound.

The key features of the algorithm are as follows.

- It usually costs $O(n^3)$ flops, but can cost up to $n^4/3$ flops if large blocks are needed (which can happen only when there are many close or clustered eigenvalues).

- It needs derivatives if there are blocks of size greater than 1. This is the price to pay for treating general f and nonnormal A (see (3.4)).

- The choice of $\delta = 0.1$ for the blocking parameter δ performs well most of time. However, it is possible for the algorithm to be unstable for all δ.

- This is the best general $f(A)$ algorithm and is the benchmark for comparing with other $f(A)$ algorithms, both general and specific.

- The algorithm is the basis of the MATLAB function funm and the NAG codes F01EK, F01EL, F01EM (real arithmetic) and F01FK, F01FL, F01FM (complex arithmetic).

6.8   (Inverse) scaling and squaring for the logarithm and exponential

The most popular approaches to computing the matrix exponential and the matrix logarithm are the (inverse) scaling and squaring methods, which employ Padé approximants $r_m \equiv r_{mm}$ together with initial transformations that ensure that the Padé approximants are sufficiently accurate. The basic identities on which the methods are based are

$$e^A \approx r_m(A/2^s)^{2^s}, \qquad r_m(x) \approx e^x,$$
$$\log(A) \approx 2^s r_m(A^{1/2^s} - I), \qquad r_m(x) \approx \log(1 + x),$$

respectively. In designing algorithms the key questions are how to choose the integers s and m. Originally a fixed choice was made, based on a priori truncation error bounds, but the state of the art algorithms use a dynamic choice that depends on A. Truncation errors from the Padé approximants are accounted for by using backward error bounds that show the approximate logarithm or exponential to be the true logarithm or exponential of a matrix within normwise relative distance u of A (these backward error bounds do not take into account rounding errors).

It is beyond our scope to describe the algorithms here. Instead we give a brief historical summary.

6.8.1 Matrix exponential

1967 The scaling and squaring method is suggested by Lawson [46].

1977 Ward [62] uses Padé degree $m = 8$ and chooses s so that $\|2^{-s}A\|_1 \leq 1$.

1978—2005 Based on error analysis of Moler and Van Loan [49], MATLAB function expm uses $m = 6$ and chooses s so that $\|2^{-s}A\|_\infty \leq 1/2$.

2005 Higham [30], [32] develops a dynamic choice of parameters allowing $m \in \{3, 5, 7, 9, 13\}$ and with $\|2^{-s}A\|_1 \leq \theta_m$ for certain parameters θ_m. This algorithm is incorporated in MATLAB R2006a.

2009 Al-Mohy and Higham [2] improve the method of Higham [30] by using sharper truncation error bounds that depend on terms $\mu_k = \|A^k\|_1^{1/k}$ instead of $\|A\|_1$. The quantities μ_k are estimated for several small values of k using a matrix norm estimator [39]. This algorithm is implemented in NAG Library routines F01EC/F01FC (Mark 25).

6.8.2 Matrix logarithm

1989 Kenney and Laub [42, App. A] introduce the inverse scaling and squaring method, used with a Schur decomposition. Square roots are computed using the Schur method of section 6.4. Based on forward error bounds in [43], Kenney and Laub take $m = 8$ and require $\|I - A^{1/2^s}\| \leq 0.25$.

1996 Dieci, Morini, and Papini [18] also use a Schur decomposition and take $m = 9$ and require $\|I - A^{1/2^s}\| \leq 0.35$.

2001 Cheng, Higham, Kenney, and Laub [13] propose a transformation-free form of the inverse scaling and squaring method that takes as a parameter the desired accuracy. Square roots are computed by the product form of the Denman–Beavers iteration (6.4), with the Padé degree m chosen dynamically. Padé approximants are evaluated using a partial fraction representation [29].

2008 Higham [31, Sec. 11.5] develops two inverse scaling and squaring algorithms, one using the Schur decomposition and one transformation-free. Both choose the Padé degree and the number of square roots dynamically. Like all previous algorithms these are based on forward error bounds for the Padé error from [43].

2012 Al-Mohy and Higham [5] develop backward error bounds expressed in terms of the quantities $\mu_k = \|A^k\|_1^{1/k}$ and incorporate them into both Schur-based and transformation-free algorithms. They also use special techniques to compute the argument $A^{1/2^s} - I$ of the Padé approximant more accurately. This work puts the inverse scaling and squaring method on par with the scaling and squaring algorithm of [2] for the matrix exponential. The Schur-based algorithm is implemented in NAG routine F01FJ (Mark 25).

2013 Al-Mohy, Higham, and Relton [6] develop a version of the algorithm of Al-Mohy and Higham [5] that works entirely in real arithmetic when the matrix is real, by exploiting the real Schur decomposition. The algorithm is implemented in NAG routine F01EJ (Mark 25).

6.9   Matrix iterations

For some matrix functions f that satisfy a nonlinear matrix equation, it is possible to derive an iteration producing a sequence X_k of matrices that converges to $f(A)$ for a suitable choice of X_0.

The practical utility of iterations for matrix functions can be destroyed by instability due to growth of errors in floating point arithmetic, so we need an appropriate definition of stability that can be used to distinguish between "good" and "bad" iterations. Let $L^i(X)$ denote the ith power of a Fréchet derivative L at X, defined as i-fold composition; thus $L^3(X, E) \equiv L(X, L(X, L(X, E)))$. Consider an iteration $X_{k+1} = g(X_k)$ with a fixed point X. Assume that g is Fréchet differentiable at X. The iteration is defined to be *stable* in a neighborhood of X if the Fréchet derivative $L_g(X)$ has bounded powers, that is, there exists a constant c such that $\|L_g^i(X)\| \leq c$ for all $i > 0$. For a stable iteration sufficiently small errors introduced near a fixed point have a bounded effect, to first order, on succeeding iterates. Note the useful standard result that a linear operator on $\mathbb{C}^{n \times n}$ is power bounded if its spectral radius is less than 1 and not power bounded if its spectral radius exceeds 1.

Let $A \in \mathbb{C}^{n \times n}$ have no pure imaginary eigenvalues and have the Jordan canonical form

$$A = Z \begin{bmatrix} J_1 & 0 \\ 0 & J_2 \end{bmatrix} Z^{-1},$$

where $J_1 \in \mathbb{C}^{p \times p}$ and $J_2 \in \mathbb{C}^{q \times q}$ have spectra in the open left half-plane and right half-plane, respectively. The *matrix sign function* is defined by

$$\operatorname{sign}(A) = Z \begin{bmatrix} -I_p & 0 \\ 0 & I_q \end{bmatrix} Z^{-1}.$$

It can be verified that the matrix sign function can also be expressed as

$$\operatorname{sign}(A) = A(A^2)^{-1/2}, \quad \operatorname{sign}(A) = \frac{2}{\pi} \int_0^\infty (t^2 I + A^2)^{-1} dt.$$

An iteration for computing the matrix sign function can be derived by applying Newton's method to $X^2 = I$:

$$X_{k+1} = \frac{1}{2}(X_k + X_k^{-1}), \qquad X_0 = A.$$

To prove convergence, let $S = \operatorname{sign}(A)$ and $G = (A - S)(A + S)^{-1}$. It can be shown that

$$X_k = (I - G^{2^k})^{-1}(I + G^{2^k})S.$$

The eigenvalues of G are of the form $(\lambda_i - \operatorname{sign}(\lambda_i))/(\lambda_i + \operatorname{sign}(\lambda_i))$, where λ_i is an eigenvalue of A. Hence $\rho(G) < 1$ and $G^k \to 0$, implying that $X_k \to S$. It is easy to show that

$$\|X_{k+1} - S\| \leq \frac{1}{2}\|X_k^{-1}\|\|X_k - S\|^2,$$

and hence the convergence is quadratic.

Consider a superlinearly convergent iteration $X_{k+1} = g(X_k)$ for $S = \operatorname{sign}(X_0)$. It can be shown that the Fréchet derivative $L_g(S, E) = \frac{1}{2}(E - SES)$, which does not depend on g. It follows that $L_g(S)$ is idempotent $(L_g^2(S) = L_g(S))$ and the iteration is stable. Hence essentially all sign iterations of practical interest are stable.

Applying Newton's method to $X^2 = A \in \mathbb{C}^{n \times n}$ yields, with X_0 given, the iteration

$$\left. \begin{array}{l} \text{Solve } X_k E_k + E_k X_k = A - X_k^2 \\ \qquad\qquad X_{k+1} = X_k + E_k \end{array} \right\} \quad k = 0, 1, 2, \ldots.$$

If X_0 commutes with A and all the iterates are defined, then this iteration simplifies to

$$X_{k+1} = \frac{1}{2}(X_k + X_k^{-1}A). \tag{6.3}$$

To analyze convergence, assume $X_0 = A$. Let $Z^{-1}AZ = J$ be a Jordan canonical form and set $Z^{-1}X_k Z = Y_k$. Then

$$Y_{k+1} = \frac{1}{2}(Y_k + Y_k^{-1}J), \qquad Y_0 = J.$$

The diagonal elements $d_i^{(k)} = (Y_k)_{ii}$ obey Heron's iteration

$$d_i^{(k+1)} = \frac{1}{2}(d_i^{(k)} + \lambda_i/d_i^{(k)}), \quad d_i^{(0)} = \lambda_i,$$

so $d_i^{(k)} \to \lambda_i^{1/2}$ assuming A has no eigenvalues on $\mathbb{R}^-$. It can also be shown that the off-diagonal elements converge and hence that $Y_k \to J^{1/2}$, or equivalently $X_k \to A^{1/2}$. However, this analysis does not generalize to general X_0 that do not commute with A.

A more general convergence result can be obtained by relating the iteration to the Newton iteration for the matrix sign function.

Theorem 6.1. *Let $A \in \mathbb{C}^{n \times n}$ have no eigenvalues on $\mathbb{R}^-$. The Newton square root iterates X_k with $X_0 A = A X_0$ are related to the Newton sign iterates*

$$S_{k+1} = \frac{1}{2}(S_k + S_k^{-1}), \qquad S_0 = A^{-1/2} X_0$$

by $X_k \equiv A^{1/2} S_k$. Hence, provided $A^{-1/2} X_0$ has no pure imaginary eigenvalues the X_k are defined and $X_k \to A^{1/2}\mathrm{sign}(S_0)$ quadratically.

The Newton iteration (6.3) is unstable, as was pointed out by Laasonen [44], who stated that

> "Newton's method if carried out indefinitely, is not stable whenever the ratio of the largest to the smallest eigenvalue of A exceeds the value 9."

Higham [27] gives analysis that explains the instability for diagonalizable A. For general A, the instability can be analyzed using our definition. The iteration function is $g(X) = (X + X^{-1}A)/2$ and its Fréchet derivative is $L_g(X, E) = (E - X^{-1}EX^{-1}A)/2$. The relevant fixed point is $X = A^{1/2}$, for which $L_g(A^{1/2}, E) = (E - A^{-1/2}EA^{1/2})/2$. The eigenvalues of $L_g(A^{1/2})$ (i.e., the eigenvalues of $(I - A^{1/2^T} \otimes A^{-1/2})/2$, where $\otimes$ denotes the Kronecker product) are $(1 - \lambda_i^{-1/2}\lambda_j^{1/2})/2$. For stability we need $\max_{i,j} \frac{1}{2}\left|1 - \lambda_i^{-1/2}\lambda_j^{1/2}\right| < 1$. For Hermitian positive definite A this reduces to the condition $\kappa_2(A) < 9$ stated by Laasonen.

Fortunately, the instability of (6.3) is not intrinsic to the method, but depends on the equations used to compute X_k. The iteration can be rewritten in various ways as a stable coupled iteration, for example as the product form of the Denman–Beavers iteration [13],

$$M_{k+1} = \frac{1}{2}\left(I + \frac{M_k + M_k^{-1}}{2}\right), \qquad M_0 = A, \qquad (6.4\text{a})$$

$$X_{k+1} = \frac{1}{2}X_k(I + M_k^{-1}), \qquad X_0 = A, \qquad (6.4\text{b})$$

for which $X_k \to A^{1/2}$ and $M_k \to I$.

7 Methods for $f(A)b$

For $A \in \mathbb{C}^{n \times n}, b \in \mathbb{C}^n$, we now consider the problem of computing $f(A)b$ without first computing $f(A)$. Cases of interest include

- $f(x) = e^x$,
- $f(x) = \log(x)$,
- $f(x) = x^\alpha$ with arbitrary real α,
- $f(x) = \text{sign}(x)$.

A minimal assumption is that matrix–vector products with A can be formed. It may also be possible to solve linear systems with A, by direct methods or iterative methods.

7.1 Krylov subspace method

A general purpose approach is to run the Arnoldi process on A and b to obtain the factorization

$$AQ_k = Q_k H_k + h_{k+1,k} q_{k+1} e_k^T,$$

where $Q_k = [q_1, q_2, \ldots, q_k]$ with $q_1 = b/\|b\|_2$ has orthonormal columns and H_k is $k \times k$ upper Hessenberg. Then we approximate

$$f(A)b \approx Q_k f(H_k) Q_k^* b = \|b\|_2 Q_k f(H_k) e_1.$$

Typically, k will be chosen relatively small in order to economize on storage ($k < 100$, say). Hence any method for dense matrices can be used for $f(H_k)e_1$. Care must be taken to guard against loss of orthogonality of Q_k in the Arnoldi process.

Among the large literature on Krylov methods for matrix functions we cite just the recent survey by Güttel [24].

7.2 $f(A)b$ via contour integration

For general f, we can use the Cauchy integral formula

$$f(A)b = \frac{1}{2\pi i} \int_\Gamma f(z)(zI - A)^{-1} b \, dz,$$

where f is analytic on and inside a closed contour Γ that encloses the spectrum of A. Assume that we can solve linear systems with A, preferably by a direct method.

We can take for the contour Γ a circle enclosing the spectrum, for example with center $(\lambda_1 + \lambda_n)/2$ and radius $\lambda_1/2$ for a symmetric positive definite matrix with spectrum in $[\lambda_n, \lambda_1]$. Then we can apply the repeated trapezium rule. However, this is inefficient unless A is very well conditioned.

Hale, Higham and Trefethen [25] use a conformal mapping, carefully constructed based on knowledge of the extreme points of the spectrum and any branch cuts or singularities of f, and then apply the repeated trapezium rule. For $A^{1/2}b$ and A with real spectrum in $[\lambda_n, \lambda_1]$ they conformally map $\mathbb{C} \setminus \{(-\infty, 0] \cup [\lambda_n, \lambda_1]\}$ to an annulus: $[\lambda_n, \lambda_1]$ is mapped to the inner circle and $(-\infty, 0]$ to the outer circle. Compared with taking a circle as described above, the conformal mapping approach reduces the number of quadrature points from 32,000 to 5 when two digits of accuracy are required or 262,000 to 35 for 13 digits.

7.3 $A^{\alpha}b$ via binomial expansion

Write $A = s(I - C)$, where we wish to choose s so that $\rho(C) < 1$. This is certainly possible if A is an M-matrix. If A has real, positive eigenvalues then $s = (\lambda_{\min} + \lambda_{\max})/2$ minimizes the spectral radius $\rho(C)$ with

$$\rho_{\min}(C) = (\lambda_{\min} - \lambda_{\max})/(\lambda_{\min} + \lambda_{\max}).$$

For any A, the value $s = \operatorname{trace}(A^*A)/\operatorname{trace}(A^*)$ minimizes $\|C\|_F$ but may or may not achieve $\rho(C) < 1$. From

$$(I - C)^{\alpha} = \sum_{j=0}^{\infty} \binom{\alpha}{j} (-C)^j, \qquad \rho(C) < 1,$$

we have

$$A^{\alpha}b = s^{\alpha} \sum_{j=0}^{\infty} \binom{\alpha}{j} (-C)^j b,$$

and this series can be truncated to approximate $A^{\alpha}b$.

7.4 $e^A b$

One of Moler and Van Loan's "nineteen dubious ways" to compute the matrix exponential [48] applies a fourth order Runge–Kutta method with fixed step size to the ODE $y' = Ay$. This produces an approximation $e^A \approx T_4(A/m)^m$, where $T_4(x)$ is a degree 4 truncation of the Taylor series for e^x. Al-Mohy and Higham [4] develop an algorithm that can be thought of as an extension of this idea to use degree s truncated Taylor series polynomials, where the degree s and the scaling parameter m are chosen to minimize the cost while ensuring a backward error of order the unit roundoff, u. It is interesting to compare this algorithm with a Krylov method (in general, Arnoldi-based):

Al-Mohy and Higham algorithm	Krylov method
Most time spent in matrix–vector products.	Cost of Krylov recurrence and computing e^H can be significant.
A "direct method", so its cost is predictable.	An iterative method; needs a stopping test.
No parameters to estimate.	Must select maximum size of Krylov subspace.
Storage: 2 vectors	Storage: Krylov basis
Can evaluate e^{At} at multiple points on the interval.	Degree of Krylov subspace will depend on t.
Works directly for $e^A B$ with a matrix B.	Need a block Krylov method for $e^A B$.
Cost tends to increase with $\|A\|$.	$\|A\|^{1/2}$-dependence of cost for symmetric negative definite A.

8   Concluding remarks

We note that this treatment is not comprehensive. For example, we have given few details about methods for the matrix exponential and matrix logarithm and not described iterative methods for pth roots [23], methods for arbitrary matrix powers [37], [38], or methods for computing Fréchet derivatives [1], [3], [6], [38].

We have also said very little about software. A catalogue of available software is given by Deadman and Higham [36].

We finish by emphasizing the importance of considering the underlying assumptions and requirements when selecting a method, as discussed in section 5. These include the desired accuracy; whether the matrix A is known explicitly or is available only in the form of a black box that returns Ax, and possibly A^*x, given x; and whether it is possible to solve $Ax = b$ by a (sparse) direct method.

References

[1] Awad H. Al-Mohy and Nicholas J. Higham. Computing the Fréchet derivative of the matrix exponential, with an application to condition number estimation. *SIAM J. Matrix Anal. Appl.*, 30(4): 1639–1657, 2009.

[2] Awad H. Al-Mohy and Nicholas J. Higham. A new scaling and squaring algorithm for the matrix exponential. *SIAM J. Matrix Anal. Appl.*, 31 (3): 970–989, 2009.

[3] Awad H. Al-Mohy and Nicholas J. Higham. The complex step approximation to the Fréchet derivative of a matrix function. *Numer. Algorithms*, 53(1): 133–148, 2010.

[4] Awad H. Al-Mohy and Nicholas J. Higham. Computing the action of the matrix exponential, with an application to exponential integrators. *SIAM J. Sci. Comput.*, 33(2): 488–511, 2011.

[5] Awad H. Al-Mohy and Nicholas J. Higham. Improved inverse scaling and squaring algorithms for the matrix logarithm. *SIAM J. Sci. Comput.*, 34 (4): C153–C169, 2012.

[6] Awad H. Al-Mohy, Nicholas J. Higham, and Samuel D. Relton. Computing the Fréchet derivative of the matrix logarithm and estimating the condition number. *SIAM J. Sci. Comput.*, 35(4): C394–C410, 2013.

[7] Mary Aprahamian and Nicholas J. Higham. The matrix unwinding function, with an application to computing the matrix exponential. *SIAM J. Matrix Anal. Appl.*, 35(1), 88–109, 2014.

[8] Zhaojun Bai and James W. Demmel. On swapping diagonal blocks in real Schur form. *Linear Algebra Appl.*, 186: 73–95, 1993.

[9] Åke Björck and Sven Hammarling. A Schur method for the square root of a matrix. *Linear Algebra Appl.*, 52/53: 127–140, 1983.

[10] A. Buchheim. An extension of a theorem of Professor Sylvester's relating to matrices. *Phil. Mag.*, 22(135): 173–174, 1886. Fifth series.

[11] Arthur Cayley. A memoir on the theory of matrices. *Philos. Trans. Roy. Soc. London*, 148: 17–37, 1858.

[12] Jie Chen, Mihai Anitescu, and Yousef Saad. Computing $f(A)b$ via least squares polynomial approximations. *SIAM J. Sci. Comput.*, 33(1): 195–222, 2011.

[13] Sheung Hun Cheng, Nicholas J. Higham, Charles S. Kenney, and Alan J. Laub. Approximating the logarithm of a matrix to specified accuracy. *SIAM J. Matrix Anal. Appl.*, 22(4): 1112–1125, 2001.

[14] M. Cipolla. Sulle matrice espressione analitiche di un'altra. *Rendiconti Circolo Matematico de Palermo*, 56: 144–154, 1932.

[15] A. R. Collar. The first fifty years of aeroelasticity. *Aerospace (Royal Aeronautical Society Journal)*, 5: 12–20, 1978.

[16] Philip I. Davies and Nicholas J. Higham. A Schur–Parlett algorithm for computing matrix functions. *SIAM J. Matrix Anal. Appl.*, 25(2): 464–485, 2003.

[17] Edvin Deadman, Nicholas J. Higham, and Rui Ralha. Blocked Schur algorithms for computing the matrix square root. In *Applied Parallel and Scientific Computing: 11th International Conference, PARA 2012, Helsinki, Finland*, P. Manninen and P. Öster, editors, volume 7782 of *Lecture Notes in Computer Science*, Springer-Verlag, Berlin, 2013, pages 171–182.

[18] Luca Dieci, Benedetta Morini, and Alessandra Papini. Computational techniques for real logarithms of matrices. *SIAM J. Matrix Anal. Appl.*, 17(3): 570–593, 1996.

[19] Ernesto Estrada and Desmond J. Higham. Network properties revealed through matrix functions. *SIAM Rev.*, 52(4): 696–714, 2010.

[20] R. A. Frazer, W. J. Duncan, and A. R. Collar. *Elementary Matrices and Some Applications to Dynamics and Differential Equations.* Cambridge University Press, 1938. xviii+416 pp. 1963 printing.

[21] G. Frobenius. Über die cogredienten Transformationen der bilinearen Formen. *Sitzungsber K. Preuss. Akad. Wiss. Berlin*, 16: 7–16, 1896.

[22] G. Giorgi. Nuove osservazioni sulle funzioni delle matrici. *Atti Accad. Lincei Rend.*, 6(8): 3–8, 1928.

[23] Chun-Hua Guo and Nicholas J. Higham. A Schur–Newton method for the matrix pth root and its inverse. *SIAM J. Matrix Anal. Appl.*, 28(3): 788–804, 2006.

[24] Stefan Güttel. Rational Krylov approximation of matrix functions: Numerical methods and optimal pole selection. *GAMM-Mitteilungen*, 36(1): 8–31, 2013.

[25] Nicholas Hale, Nicholas J. Higham, and Lloyd N. Trefethen. Computing A^{α}, $\log(A)$, and related matrix functions by contour integrals. *SIAM J. Numer. Anal.*, 46(5): 2505–2523, 2008.

[26] W. F. Harris. The average eye. *Opthal. Physiol. Opt.*, 24: 580–585, 2005.

[27] Nicholas J. Higham. Newton's method for the matrix square root. *Math. Comp.*, 46(174): 537–549, 1986.

[28] Nicholas J. Higham. Computing real square roots of a real matrix. *Linear Algebra Appl.*, 88/89: 405–430, 1987.

[29] Nicholas J. Higham. Evaluating Padé approximants of the matrix logarithm. *SIAM J. Matrix Anal. Appl.*, 22(4): 1126–1135, 2001.

[30] Nicholas J. Higham. The scaling and squaring method for the matrix exponential revisited. *SIAM J. Matrix Anal. Appl.*, 26(4): 1179–1193, 2005.

[31] Nicholas J. Higham. *Functions of Matrices: Theory and Computation.* Society for Industrial and Applied Mathematics, Philadelphia, PA, USA, 2008. xx+425 pp. ISBN 978-0-898716-46-7.

[32] Nicholas J. Higham. The scaling and squaring method for the matrix exponential revisited. *SIAM Rev.*, 51(4): 747–764, 2009.

[33] Nicholas J. Higham. Arthur Buchheim (1859–1888). http://nickhigham.wordpress.com/2013/01/31/arthur-buchheim/, 2013.

[34] Nicholas J. Higham. Gene Golub SIAM Summer School 2013. http://nickhigham.wordpress.com/2013/08/09/gene-golub-siam -summer-school-2013/, 2013.

[35] Nicholas J. Higham and Awad H. Al-Mohy. Computing matrix functions. *Acta Numerica*, 19: 159–208, 2010.

[36] Nicholas J. Higham and Edvin Deadman. A catalogue of software for matrix functions. Version 1.0. MIMS EPrint 2014.8, Manchester Institute for Mathematical Sciences, The University of Manchester, UK, February 2014. 19 pp.

[37] Nicholas J. Higham and Lijing Lin. A Schur–Padé algorithm for fractional powers of a matrix. *SIAM J. Matrix Anal. Appl.*, 32(3): 1056–1078, 2011.

[38] Nicholas J. Higham and Lijing Lin. An improved Schur–Padé algorithm for fractional powers of a matrix and their Fréchet derivatives. *SIAM J. Matrix Anal. Appl.*, 34(3): 1341–1360, 2013.

[39] Nicholas J. Higham and Françoise Tisseur. A block algorithm for matrix 1-norm estimation, with an application to 1-norm pseudospectra. *SIAM J. Matrix Anal. Appl.*, 21(4): 1185–1201, 2000.

[40] Marlis Hochbruck and Alexander Ostermann. Exponential integrators. *Acta Numerica*, 19: 209–286, 2010.

[41] Leo Katz. A new status index derived from sociometric analysis. *Psychometrika*, 18(1): 39–43, 1953.

[42] Charles S. Kenney and Alan J. Laub. Condition estimates for matrix functions. *SIAM J. Matrix Anal. Appl.*, 10(2): 191–209, 1989.

[43] Charles S. Kenney and Alan J. Laub. Padé error estimates for the logarithm of a matrix. *Internat. J. Control*, 50(3): 707–730, 1989.

[44] Pentti Laasonen. On the iterative solution of the matrix equation $AX^2 - I = 0$. *M.T.A.C.*, 12: 109–116, 1958.

[45] Edmond Nicolas Laguerre. Le calcul des systèmes linéaires, extrait d'une lettre adressé à M. Hermite. In *Oeuvres de Laguerre*, Ch. Hermite, H. Poincaré, and E. Rouché, editors, volume 1, Gauthier–Villars, Paris, 1898, pages 221–267. The article is dated 1867 and is "Extrait du Journal de l'École Polytechnique, LXIIᵉ Cahier".

[46] J. Douglas Lawson. Generalized Runge-Kutta processes for stable systems with large Lipschitz constants. *SIAM J. Numer. Anal.*, 4(3): 372–380, 1967.

[47] Borislav V. Minchev and Will M. Wright. A review of exponential integrators for first order semi-linear problems. Preprint 2/2005, Norwegian University of Science and Technology, Trondheim, Norway, 2005. 44 pp.

[48] Cleve B. Moler and Charles F. Van Loan. Nineteen dubious ways to compute the exponential of a matrix, twenty-five years later. *SIAM Rev.*, 45(1): 3–49, 2003.

[49] Cleve B. Moler and Charles F. Van Loan. Nineteen dubious ways to compute the exponential of a matrix. *SIAM Rev.*, 20(4): 801–836, 1978.

[50] Beresford N. Parlett. Computation of functions of triangular matrices. Memorandum ERL-M481, Electronics Research Laboratory, College of Engineering, University of California, Berkeley, November 1974. 18 pp.

[51] Beresford N. Parlett. A recurrence among the elements of functions of triangular matrices. *Linear Algebra Appl.*, 14: 117–121, 1976.

[52] Michael S. Paterson and Larry J. Stockmeyer. On the number of nonscalar multiplications necessary to evaluate polynomials. *SIAM J. Comput.*, 2 (1): 60–66, 1973.

[53] G. Peano. Intégration par Séries des équations différentielles linéaires. *Math. Annalen*, 32: 450–456, 1888.

[54] H. Poincaré. Sur les groupes continus. *Trans. Cambridge Phil. Soc.*, 18: 220–255, 1899.

[55] R. F. Rinehart. The equivalence of definitions of a matric function. *Amer. Math. Monthly*, 62: 395–414, 1955.

[56] Hans Schwerdtfeger. *Les Fonctions de Matrices. I. Les Fonctions Univalentes*. Number 649 in *Actualités Scientifiques et Industrielles*. Hermann, Paris, France, 1938. 58 pp.

[57] Matthew I. Smith. A Schur algorithm for computing matrix pth roots. *SIAM J. Matrix Anal. Appl.*, 24(4): 971–989, 2003.

[58] J. J. Sylvester. Additions to the articles, "On a New Class of Theorems", and "On Pascal's Theorem". *Philosophical Magazine*, 37: 363–370, 1850. Reprinted in [60, pp. 145–151].

[59] J. J. Sylvester. On the equation to the secular inequalities in the planetary theory. *Philosophical Magazine*, 16: 267–269, 1883. Reprinted in [61, pp. 110–111].

[60] *The Collected Mathematical Papers of James Joseph Sylvester*, volume 1 (1837—1853). Cambridge University Press, 1904. xii+650 pp.

[61] *The Collected Mathematical Papers of James Joseph Sylvester*, volume IV (1882—1897). Chelsea, New York, 1973. xxxvii+756 pp. ISBN 0-8284-0253-1.

[62] Robert C. Ward. Numerical computation of the matrix exponential with accuracy estimate. *SIAM J. Numer. Anal.*, 14(4): 600–610, 1977.

A Short Course on Exponential Integrators

Marlis Hochbruck*

Abstract

This paper contains a short course on the construction, analysis, and implementation of exponential integrators for time dependent partial differential equations. A much more detailed recent review can be found in Hochbruck and Ostermann (2010). Here, we restrict ourselves to one-step methods for autonomous problems.

A basic principle for the construction of exponential integrators is the linearization of a semilinear or a nonlinear evolution equation. We distinguish exponential Runge–Kutta methods, using a fixed linearization and exponential Rosenbrock-type methods, which use a continuous linearization at the current approximation of the solution. We present some of the convergence results and give a proof for the simplest method, the exponential Euler method.

The fact that it is possible to construct explicit exponential integrators which obey error bounds even for abstract evolution equations comes at the price that one has to approximate products of matrix functions with vectors in the spatially discrete case. For an efficient implementation one has to combine the integrator with well-chosen algorithms from numerical linear algebra. We briefly sketch Krylov subspace methods for this task.

1 Construction

In this section we will discuss methods for autonomous evolution equations of the form

$$u'(t) = F\big(u(t)\big), \qquad u(t_0) = u_0, \tag{1.1}$$

in a finite time interval $t \in [t_0, T]$. Our main interest is in the case where F is a partial differential operator or its spatial discretization. Note that in the abstract case, F involves an unbounded operator and

---

*Department of Mathematics, Karlsruhe Institute of Technology, Germany (marlis.hochbruck@kit.edu).

for the discrete case, the norm of the Jacobian grows with the inverse of the spatial mesh width.

For a recent survey on exponential integrators see Hochbruck and Ostermann (2010).

1.1   Runge–Kutta methods

Before we start to introduce exponential integrators, let us briefly review Runge–Kutta methods for the solution of the ordinary differential equation

$$u'(t) = g\big(u(t)\big), \qquad u(t_0) = u_0. \tag{1.2}$$

The function g is assumed to satisfy a Lipschitz condition with a moderate Lipschitz constant (in contrast to F in the evolution equation (1.1)).

The construction of Runge–Kutta methods relies on the following representation of the exact solution u of (1.2) at time $t_{n+1} = t_n + \tau$,

$$u(t_{n+1}) = u(t_n) + \int_0^\tau g\big(u(t_n + \theta)\big)d\theta, \qquad n = 0, 1, \dots. \tag{1.3}$$

The idea is to approximate the integral on the right-hand side by a quadrature formula defined by nodes $0 \le c_1 < \cdots < c_s \le 1$ and weights $b_1, \dots, b_s$. Assume we are given approximations $u_n \approx u(t_n)$ and

$$U_{ni} \approx u(t_n + c_i\tau).$$

Then we have

$$u(t_{n+1}) \approx u_n + \tau \sum_{i=1}^s b_i U'_{ni}, \quad \text{where} \quad U'_{ni} = g\big(U_{ni}\big), \quad i = 1, \dots, s.$$

The approximations U_{ni}, $i = 1, \dots, s$, can be obtained by yet other quadrature formulas via

$$u(t_n + c_i\tau) = u(t_n) + \int_0^{c_i\tau} u'(t_n + \theta)d\theta \approx u_n + \tau \sum_{j=1}^s a_{ij} U'_{nj}.$$

A general s-stage Runge–Kutta method is defined as

$$u_{n+1} = u_n + \tau \sum_{i=1}^s b_i U'_{ni},$$

$$U'_{ni} = g(U_{ni}), \qquad\qquad i = 1, \dots, s, \tag{1.4}$$

$$U_{ni} = u_n + \tau \sum_{j=1}^s a_{ij} U'_{nj}, \qquad i = 1, \dots, s,$$

and denoted in a so-called Butcher tableau as

$$\frac{c_i \mid a_{ij}}{\mid b_j}$$

If $a_{ij} = 0$ for $j \geq i$, then U_{ni}, $i = 1, \ldots, s$, can be computed explicitly by just evaluating the function g at already computed approximations. Hence, these methods are called explicit Runge–Kutta methods.

Runge–Kutta methods belong to the class of one-step methods, since they only use the current approximation $u_n \approx u(t_n)$ to construct the new approximation $u_{n+1} \approx u(t_n + \tau)$.

Definition 1.1. *The local error of a one-step method for solving the initial value problem* (1.2) *is defined as*

$$u_1 - u(t_0 + \tau),$$

where u_1 is the approximation obtained from $u_0 = u(t_0)$ after one step with step size τ.

An important property of methods for solving initial value problems is the order.

Definition 1.2. *A numerical scheme for solving the initial value problem* (1.2) *is of* **order** *p if for any $g \in C^{p+1}$ the local error is of size $\mathcal{O}(\tau^{p+1})$. If $p \geq 1$, the method is called* **consistent**.

When using the notation $\mathcal{O}$, it is of utmost importance to mention all assumptions and quantities which enter the constant defining the set. This is particularly true in the analysis of time-dependent partial differential equations, no matter if the abstract problem is considered in a function space or if a finite difference, a finite element, or a spectral method is used for the spatial discretization.

To verify that a Runge–Kutta method is of a certain order, one has to compute the Taylor expansion of the numerical solution and the Taylor expansion of the exact solution. This requires assumptions on the smoothness of the exact solution (namely that all derivatives used within the expansion are bounded in a suitable norm).

Example 1.3. The simplest method is the well-known explicit Euler method

$$u_{n+1} = u_n + \tau g(u_n), \qquad \text{Butcher tableau} \qquad \frac{0 \mid 0}{\mid 1}$$

which is of order one.

1.2   Exponential Runge–Kutta methods

Next we study the construction of exponential Runge–Kutta methods. Such methods are based on linearizing F in (1.1). There are two main options: the first one uses a fixed linearization

$$F\big(u(t)\big) = -Au(t) + g\big(u(t)\big), \qquad A \approx -\frac{\partial F}{\partial u}(u_0). \qquad (1.5)$$

The second option is based on a continuous linearization around the current approximation $u_n \approx u(t_n)$

$$\begin{aligned} F\big(u(t)\big) &= -A_n u(t) + g_n\big(u(t)\big), \\ A_n &= -\frac{\partial F}{\partial u}(u_n), \end{aligned} \qquad n = 0, 1, \dots. \qquad (1.6)$$

We start with one-step methods based on the fixed linearization (1.5) and consider the initial value problem in a finite dimensional space (say $\mathbb{R}^d$ or $\mathbb{C}^d$). The construction of *exponential Runge–Kutta methods* is closely related to the construction of standard Runge–Kutta methods. It relies on the variation-of-constants formula

$$u(t_n + \tau) = e^{-\tau A} u(t_n) + \int_0^\tau e^{-(\tau - \sigma)A} g\big(u(t_n + \sigma)\big) d\sigma. \qquad (1.7)$$

Here, we used the notation $e^{-\tau A}$ for the matrix exponential or, in a functional analytic framework, for a semigroup generated by $-A$.

We choose nodes $0 \le c_1 < \dots < c_s \le 1$ and assume we are given approximations

$$U_{ni} \approx u(t_n + c_i \tau), \qquad i = 1, \dots, s. \qquad (1.8)$$

The integral in (1.7) is approximated by a quadrature formula, in which only the nonlinearity g is approximated but the exponential operator (semigroup) is treated exactly:

$$u_{n+1} = e^{-\tau A} u_n + \tau \sum_{i=1}^{s} b_i(-\tau A) G_{ni}, \qquad (1.9a)$$

$$G_{ni} = g\big(U_{ni}\big), \qquad (1.9b)$$

$$U_{ni} = e^{-c_i \tau A} u_n + \tau \sum_{j=1}^{s} a_{ij}(-\tau A) G_{nj}. \qquad (1.9c)$$

Note that choosing $A = 0$, *i.e.*, $F = g$, the method reduces to a standard Runge–Kutta method (1.4) with coefficients $b_i(0)$, $a_{ij}(0)$. This Runge–Kutta method will be called the *underlying Runge–Kutta method*. We can gather the exponential Runge–Kutta method in the following tableau

$$\frac{c_i \;\big|\; a_{ij}(z)}{\big|\; b_j(z)}, \tag{1.10}$$

where the coefficients are analytic functions which are evaluated at the linear operator $-\tau A$.

Since we use the exponential function within the integrator, a natural requirement is to enforce the integrator to solve linear problems (1.1) with *constant* g exactly. Then

$$u(t_n + \theta\tau) = e^{-\theta\tau A}u(t_n) + \theta\tau\varphi_1(-\theta\tau A)g, \tag{1.11}$$

where

$$\varphi_1(z) = \int_0^1 e^{(1-\sigma)z}d\sigma = \frac{e^z - 1}{z}. \tag{1.12}$$

We use the representation (1.11) for $\theta = c_i$, $i = 1, \ldots, s$, and $\theta = 1$. The conditions

$$u_n = u(t_n), \quad U_{ni} = u(t_n + c_i\tau), \quad n = 0, 1, \ldots, \; i = 1, \ldots, s \tag{1.13}$$

are fulfilled if the *simplifying assumptions*

$$\sum_{i=1}^s b_i(z) = \varphi_1(z), \tag{1.14a}$$

$$\sum_{j=1}^s a_{ij}(z) = c_i\varphi_1(c_iz), \qquad i = 1, \ldots, s \tag{1.14b}$$

are satisfied. We restrict ourselves to explicit methods, where $a_{ij}(z) = 0$ for $i \leq j$.

Remark. For $z = 0$, the simplifying assumptions just give

$$\sum_{i=1}^s b_i(0) = 1, \qquad \sum_{j=1}^s a_{ij}(0) = c_i,$$

which ensures that the underlying Runge–Kutta method has stage order $q \geq 1$ (*i.e.*, all inner quadrature formulas are of order at least one) and order $p \geq 1$.

Example 1.4. For $s = 1$, (1.14) yields $b_1(z) = \varphi_1(z)$. The resulting method is called *exponential Euler method*. By $e^{-z} = 1 - z\varphi_1(-z)$ we have the following equivalent representations of this scheme

$$\begin{aligned}u_{n+1} &= e^{-\tau A}u_n + \tau\varphi_1(-\tau A)g(u_n)\\ &= (I - \tau A\varphi_1(-\tau A))u_n + \tau\varphi_1(-\tau A)g(u_n)\\ &= u_n + \tau\varphi_1(-\tau A)F(u_n).\end{aligned} \tag{1.15}$$

For a practical application, where A is a large scale matrix stemming from the space discretization of a differential operator, the latter formula is computationally more efficient, since it requires the evaluation of one product of a matrix function $\varphi_1(-\tau A)$ times a vector, while the first formula requires two such evaluations.

The computation of one time step of the general scheme (1.9) with $s > 1$ requires evaluations of the product of matrix functions with $s + 1$ different vectors, namely with u_n and G_{ni}, $i = 1, \ldots, s$. An attractive option to evaluate these products is using an iterative method (for instance a Krylov subspace method with respect to the matrix A and each of the $s + 1$ vectors). We will see later that it is advantageous to apply the iterative method to vectors of small norm. This can be achieved by defining

$$D_{ni} = G_{ni} - g(u_n), \qquad i = 1, \ldots, s. \tag{1.16}$$

Note that for explicit methods, $D_{n1} = 0$, and $\|D_{ni}\| = \mathcal{O}(\tau)$, $i = 2, \ldots, s$.

Using the simplifying assumptions (1.14), we can reformulate the method (1.9) equivalently as

$$U_{ni} = u_n + c_i \tau \varphi_1(-c_i \tau A) F(u_n) + \tau \sum_{j=2}^{i-1} a_{ij}(-\tau A) D_{nj}, \tag{1.17a}$$

$$D_{ni} = g(U_{ni}) - g(u_n), \tag{1.17b}$$

$$u_{n+1} = u_n + \tau \varphi_1(-\tau A) F(u_n) + \tau \sum_{i=2}^{s} b_i(-\tau A) D_{ni}. \tag{1.17c}$$

Note that the inner stages U_{ni} and u_{n+1} can be interpreted as corrected exponential Euler approximations with step size $c_i \tau$ and τ, respectively.

The implementation of this method requires the evaluation of the product of matrix functions with s vectors $F(u_n)$, and D_{nj}, $j = 2, \ldots, s$. Only the vector $F(u_n)$ is of norm $\mathcal{O}(1)$, but the remaining $s - 1$ vectors D_{nj} are of size $\mathcal{O}(\tau)$, so we can hope that only one expensive approximation is required.

A MATLAB software package with an implementation of exponential Runge–Kutta methods and exponential multistep methods for test problems is provided by Berland, Skaflestad, and Wright (2007).

1.3   Exponential Rosenbrock-type methods

These methods are based on the continuous linearization (1.6) of the differential equation (1.1). The idea is similar to that of exponential Runge–Kutta methods, namely we start from the variation-of-constants formula and approximate only g_n but not the exponential operators.

We restrict ourselves to the more interesting case of explicit methods satisfying the simplifying assumptions (1.14).

The variation-of-constants formula (1.7) for the solution of

$$u' = F(u) = -A_n u(t) + g_n\big(u(t)\big)$$

yields

$$u(t_n + \theta\tau) = e^{-\theta\tau A_n} u(t_n) + \int_0^{\theta\tau} e^{-(\theta\tau-\sigma)A_n} g_n\big(u(t_n+\sigma)\big)d\sigma.$$

We use this representation for $\theta = c_i$, $i = 1,\ldots,s$, and $\theta = 1$. With

$$D_{ni} = g_n(U_{ni}) - g_n(u_n), \tag{1.18a}$$

we define approximations $U_{ni} \approx u(t_n + c_i\tau)$ by

$$U_{ni} = e^{-c_i\tau A_n} u_n + c_i\tau\varphi_1(-c_i\tau A_n)g_n(u_n) + \tau\sum_{j=2}^{i-1} a_{ij}(-\tau A_n)D_{nj}$$

$$= u_n + c_i\tau\varphi_1(-c_i\tau A_n)F(u_n) + \tau\sum_{j=2}^{i-1} a_{ij}(-\tau A_n)D_{nj}, \tag{1.18b}$$

and

$$u_{n+1} = u_n + \tau\varphi_1(-\tau A_n)F(u_n) + \tau\sum_{i=2}^{s} b_i(-\tau A_n)D_{ni}. \tag{1.18c}$$

Example 1.5. Since $c_1 = 0$, for $s = 1$ we obtain the exponential Rosenbrock–Euler method proposed by Pope (1963).

$$u_{n+1} = e^{-\tau A_n} u_n + \tau\varphi_1(-\tau A_n)g_n(u_n)$$
$$= u_n + \tau\varphi_1(-\tau A_n)F(u_n). \tag{1.19}$$

It requires only one matrix function per step.

Clearly, as for exponential Runge–Kutta methods, the approximations U_{ni} and u_{n+1} in (1.18) can be interpreted as corrections of the exponential Rosenbrock-Euler method.

2 Error bounds

In this section we present some basic ideas on how to derive error bounds for exponential integrators for partial differential equations. We treat abstract evolution equations and their spatial semidiscretization in a uniform functional analytical framework. This leads to bounds which are independent of the spatial mesh width.

2.1 Analytical framework

We study exponential integrators in a framework of semigroups, see Engel and Nagel (2006) for a short course.

Assumption 2.1. *Let X be a Banach space with norm $\|\cdot\|$. We assume that A is a linear operator on X and that $(-A)$ is the infinitesimal generator of a strongly continuous semigroup e^{-tA} on X.*

This assumption implies

$$\|e^{-tA}\|_{X \leftarrow X} \leq C_A\, e^{-\omega t}, \quad t \geq 0 \tag{2.1}$$

with $C_A \geq 1$, $\omega \in \mathbb{R}$. We will use only this bound for our error analysis.

For this lecture, the situation in the following example is the relevant one and we just stated the assumption in this generality for people familiar with semigroup theory.

Example 2.2. For $X = \mathbb{R}^n$ or $X = \mathbb{C}^n$, the operator A can be represented by an $n \times n$ matrix $\mathbf{A}$ and $e^{-t\mathbf{A}}$ is just the matrix exponential function. For the spectral norm, the bound (2.1) is satisfied with $C_{\mathbf{A}} = 1$ if the field of values of $\mathbf{A}$ is contained in the complex half-plane

$$\mathbb{C}_\omega := \{z \in \mathbb{C} \colon \operatorname{Re} z \geq \omega\}.$$

In this case, $\omega = -\mu(-\mathbf{A})$, where $\mu(\mathbf{B}) = \lambda_{\max}\left(\frac{1}{2}\left(\mathbf{B} + \mathbf{B}^H\right)\right)$ is the so-called *logarithmic norm* of the matrix $\mathbf{B}$.

As a special case, the bound (2.1) holds with $\omega = 0$ if the field of values $\mathcal{F}(\mathbf{A})$ of $\mathbf{A}$ defined as

$$\mathcal{F}(\mathbf{A}) = \{\mathbf{x}^H \mathbf{A}\mathbf{x}, \ \|\mathbf{x}\| = 1\}$$

is contained in the right complex half-plane. In particular, this is true if $\mathbf{A} = \mathbf{A}^H$ is positive semidefinite or $\mathbf{A} = -\mathbf{A}^H$.

If $\mathbf{A}$ is diagonalizable, $\mathbf{X}^{-1}\mathbf{A}\mathbf{X} = \mathbf{\Lambda}$, then (2.1) holds for an arbitrary matrix norm induced by a vector norm with $C_{\mathbf{A}} = \kappa(\mathbf{X}) = \|\mathbf{X}\|\,\|\mathbf{X}^{-1}\|$ if the spectrum of $\mathbf{A}$ is contained in $\mathbb{C}_\omega$.

The crucial observation is that these assumptions are independent of the dimension of $\mathbf{A}$. This will allow one to prove temporal convergence results that are independent of the spatial mesh.

2.2 Exponential Runge–Kutta methods: analysis

Our analysis will make use of the variation-of-constants formula (1.7) for the solution of

$$u'(t) + Au(t) = g\big(u(t)\big), \quad u(t_0) = u_0, \tag{2.2}$$

i.e., (1.1) with fixed linearization (1.5). In order to simplify the notation, we set

$$f(t) = g\big(u(t)\big).$$

For the nonlinearity g we make the following assumption (for an analysis under more general assumptions we refer to Hochbruck and Ostermann (2005)):

Assumption 2.3. *We assume that $g : X \to X$ is locally Lipschitz-continuous in a strip along the exact solution u. Thus, there exists a real number $L = L(R,T)$ such that, for all $t \in [0,T]$,*

$$\|g(v) - g(w)\| \leq L\|v - w\| \tag{2.3}$$

if $\max\big(\|v - u(t)\|, \|w - u(t)\|\big) \leq R$.

Our proofs are heavily based on the representation of the exact solution by the variation-of-constants formula (1.7), which coincides with

$$u(t_{n+1}) = e^{-\tau A}u(t_n) + \int_0^\tau e^{-(\tau-\sigma)A}f(t_n + \sigma)d\sigma \tag{2.4}$$

in our notation.

In order to analyze exponential Runge–Kutta methods, we expand $f(t)$ in a Taylor series with remainder in integral form and insert it into (2.4):

$$u(t_{n+1}) = e^{-\tau A}u(t_n) + \int_0^\tau e^{-(\tau-\sigma)A}f(t_n + \sigma)d\sigma$$

$$= e^{-\tau A}u(t_n) + \tau \sum_{k=1}^p \varphi_k(-\tau A)\tau^{k-1}f^{(k-1)}(t_n) \tag{2.5}$$

$$+ \int_0^\tau e^{-(\tau-\sigma)A}\int_0^\sigma \frac{(\sigma-\xi)^{p-1}}{(p-1)!} f^{(p)}(t_n + \xi)d\xi d\sigma.$$

Here we used the φ-functions defined as

$$\varphi_k(z) = \int_0^1 e^{(1-\theta)z} \frac{\theta^{k-1}}{(k-1)!}d\theta, \quad k \geq 1. \tag{2.6}$$

These functions satisfy $\varphi_k(0) = 1/k!$ and the recurrence relation

$$\varphi_{k+1}(z) = \frac{\varphi_k(z) - \varphi_k(0)}{z}, \quad \varphi_0(z) = e^z. \tag{2.7}$$

Assumption 2.1 enables us to define the operators

$$\varphi_k(-\tau A) = \int_0^1 e^{-\tau(1-\theta)A} \frac{\theta^{k-1}}{(k-1)!}d\theta, \quad k \geq 1.$$

The following lemma turns out to be crucial.

Lemma 2.4. *Under Assumption* 2.1, *the operators* $\varphi_k(-\tau A)$, $k = 1$, $2,\ldots$, *are bounded on* X.

Proof. The boundedness simply follows from the estimate

$$\|\varphi_k(-\tau A)\|_{X \leftarrow X} \leq \int_0^1 \|e^{-\tau(1-\theta)A}\|_{X \leftarrow X} \frac{\theta^{k-1}}{(k-1)!} d\theta$$

and the bound (2.1) on the semigroup. $\qquad\square$

We now present an error bound for the simplest exponential integrator, the exponential Euler method.

The Taylor expansion of f with remainder in integral form is given by

$$f(t_n + \sigma) = f(t_n) + \int_0^\sigma f'(t_n + \theta)d\theta. \tag{2.8}$$

Inserting the exact solution into the numerical scheme yields

$$u(t_{n+1}) = e^{-\tau A}u(t_n) + \tau\varphi_1(-\tau A)f(t_n) + \delta_{n+1}, \tag{2.9}$$

where, by (2.5), the defect is given by

$$\delta_{n+1} = \int_0^\tau e^{-(\tau-\sigma)A} \int_0^\sigma f'(t_n + \theta)d\theta d\sigma. \tag{2.10}$$

For this defect we have the following estimate.

Lemma 2.5. *Let the semilinear initial value problem satisfy Assumption* 2.1 *and assume that* $f' \in L^\infty(0, T; X)$. *Then*

$$\left\|\sum_{j=0}^{n-1} e^{-j\tau A}\delta_{n-j}\right\| \leq C\tau M, \qquad M := \sup_{0 \leq t \leq t_n} \|f'(t)\| \tag{2.11}$$

holds with a constant $C = C(C_A, \omega, t_n)$, *uniformly in* $0 \leq t_n \leq T$.

Proof. We write

$$e^{-j\tau A}\delta_{n-j} = e^{-j\tau A} \int_0^\tau e^{-(\tau-\sigma)A} \int_0^\sigma f'(t_{n-j-1} + \theta)d\theta d\sigma.$$

Using the stability bound (2.1), the sum is bounded by

$$\left\|\sum_{j=0}^{n-1} e^{-j\tau A}\delta_{n-j}\right\| \leq CM\tau^2 n \leq CM\tau t_n.$$

This proves the desired estimate. $\qquad\square$

For the exponential Euler method, we have the following convergence result.

Theorem 2.6. *Let the initial value problem* (2.2) *satisfy Assumption* 2.1, *and consider for its numerical solution the exponential Euler method* (1.15). *Further assume that $f : [0, T] \to X$ is differentiable with $f' \in L^\infty(0, T; X)$. Then, the error bound*

$$\|u_n - u(t_n)\| \leq C\tau \sup_{0 \leq t \leq t_n} \|f'(t)\|$$

holds uniformly in $0 \leq n\tau \leq T$. The constant C depends on T, but it is independent of n and τ.

Proof. Let the error be denoted by $e_n = u_n - u(t_n)$. Then the exponential Euler method satisfies the error recursion

$$e_{n+1} = e^{-\tau A}e_n + \tau\varphi_1(-\tau A)\big(g(u_n) - f(t_n)\big) - \delta_{n+1} \qquad (2.12)$$

with defect δ_{n+1} defined in (2.10). Solving this recursion yields

$$e_n = \tau \sum_{j=0}^{n-1} e^{-(n-j-1)\tau A}\varphi_1(-\tau A)\big(g(u_j) - f(t_j)\big) - \sum_{j=0}^{n-1} e^{-j\tau A}\delta_{n-j}.$$

Using (2.1) and Lemma 2.5, we may estimate this by

$$\|e_n\| \leq C\tau \sum_{j=0}^{n-1} \|e_j\| + C\tau \sup_{0 \leq t \leq t_n} \|f'(t)\|.$$

The application of the discrete Gronwall Lemma 2.7 concludes the proof. □

In the previous proof we used the following standard discrete Gronwall Lemma.

Lemma 2.7 (Discrete Gronwall Lemma). *For $\tau > 0$ and $T > 0$, let $0 \leq t_n = n\tau \leq T$. Further assume that the sequence of non-negative numbers ε_n satisfies the inequality*

$$\varepsilon_n \leq a\tau \sum_{\nu=1}^{n-1} \varepsilon_\nu + b$$

for some $a, b \geq 0$. Then the estimate $\varepsilon_n \leq Cb$ holds, where the constant C depends on a and T.

The convergence analysis of higher-order methods turns out to be much more complicated than that for the exponential Euler scheme, due to the low order of the internal stages. The order conditions in Table 1 contain the functions

$$\psi_j(-\tau A) = \varphi_j(-\tau A) - \sum_{i=1}^{s} b_i(-\tau A) \frac{c_i^{j-1}}{(j-1)!} \qquad (2.13)$$

and

$$\psi_{j,i}(-\tau A) = \varphi_j(-c_i\tau A)c_i^j - \sum_{k=1}^{i-1} a_{ik}(-\tau A) \frac{c_k^{j-1}}{(j-1)!} \qquad (2.14)$$

which arise in the Taylor expansion within the variation-of-constants formula.

Theorem 2.8 (Hochbruck and Ostermann (2005)). *Let the initial value problem* (2.2) *satisfy Assumptions* 2.9 *and* 2.10 *and consider for its numerical solution an explicit exponential Runge–Kutta method* (1.9) *satisfying* (1.14). *For* $2 \leq p \leq 4$, *assume that the order conditions of Table 1 hold up to order* $p-1$ *and that* $\psi_p(0) = 0$. *Further assume that the remaining conditions of order p hold in a weaker form with* $b_i(0)$ *instead of* $b_i(-\tau A)$ *for* $2 \leq i \leq s$. *Then the numerical solution* u_n *satisfies the error bound*

$$\|u_n - u(t_n)\| \leq C\,\tau^p$$

uniformly in $0 \leq n\tau \leq T$. *The constant C depends on T, but it is independent of n and* τ.

Table 1   Stiff order conditions for explicit exponential Runge–Kutta methods. Here J and K denote arbitrary bounded operators on X. The functions ψ_i and $\psi_{k,\ell}$ are defined in (2.13) and (2.14), respectively.

Number	Order	Order condition
1	1	$\psi_1(-\tau A) = 0$
2	2	$\psi_2(-\tau A) = 0$
3	2	$\psi_{1,i}(-\tau A) = 0$
4	3	$\psi_3(-\tau A) = 0$
5	3	$\sum_{i=1}^{s} b_i(-\tau A)J\psi_{2,i}(-\tau A) = 0$
6	4	$\psi_4(-\tau A) = 0$
7	4	$\sum_{i=1}^{s} b_i(-\tau A)J\psi_{3,i}(-\tau A) = 0$
8	4	$\sum_{i=1}^{s} b_i(-\tau A)J\sum_{j=2}^{i-1} a_{ij}(-\tau A)J\psi_{2,j}(-\tau A) = 0$
9	4	$\sum_{i=1}^{s} b_i(-\tau A)c_iK\psi_{2,i}(-\tau A) = 0$

Examples of higher order methods and references to them can be found in Hochbruck and Ostermann (2005), Hochbruck and Ostermann (2010), and Luan and Ostermann (2014a).

2.3   Exponential Rosenbrock-type methods: analysis

For the error analysis of (1.18), we work in a semigroup framework. Background information on semigroups can be found in the textbooks Engel and Nagel (2000); Pazy (1992). Let

$$J = J(u) = \mathrm{D}F(u) = \frac{\partial F}{\partial u}(u) \qquad (2.15)$$

be the Fréchet derivative of F in a neighborhood of the exact solution of (1.5). Throughout the paper we consider the following assumptions.

Assumption 2.9. *The linear operator $J = J(u)$ is the generator of a strongly continuous semigroup e^{tJ} on a Banach space X. More precisely, we assume that there exist constants C and ω such that*

$$\left\| e^{tJ} \right\|_{X \leftarrow X} \leq C \, e^{\omega t}, \qquad t \geq 0 \qquad (2.16)$$

holds uniformly in a neighborhood of the exact solution of (1.5).

In the subsequent analysis, we restrict our attention to autonomous semilinear problems,

$$u'(t) = F\big(u(t)\big), \quad F(u) = -Au + g(u), \quad u(t_0) = u_0. \qquad (2.17)$$

This implies that (1.6) takes the form

$$- A_n = -A + \frac{\partial g}{\partial u}(u_n), \quad g_n\big(u(t)\big) = g\big(u(t)\big) - \frac{\partial g}{\partial u}(u_n)u(t). \qquad (2.18)$$

We suppose that A satisfies Assumption 2.1. Our main hypothesis on the nonlinearity g is the following:

Assumption 2.10. *We assume that (2.2) possesses a sufficiently smooth solution $u : [0, T] \to X$ with derivatives in X, and that $g : X \to X$ is sufficiently often Fréchet-differentiable in a strip along the exact solution. All occurring derivatives are assumed to be uniformly bounded.*

The latter assumption implies that the Jacobian (2.15) satisfies the Lipschitz condition

$$\left\| J(u) - J(v) \right\|_{X \leftarrow X} \leq C \left\| u - v \right\| \qquad (2.19)$$

in a neighborhood of the exact solution.

Theorem 2.11 (Hochbruck et al., 2009, Theorem 4.1). *Suppose the initial value problem* (2.17) *satisfies Assumptions* 2.1 *and* 2.10. *Consider for its numerical solution an explicit exponential Rosenbrock method* (1.18) *that fulfills the order conditions of Table 2 up to order p for some* $2 \le p \le 4$. *Further, let the step size sequence* τ_j *satisfy the condition*

$$\sum_{k=1}^{n-1}\sum_{j=0}^{k-1} \tau_j^{p+1} \le C_\tau \tag{2.20}$$

with a constant C_τ *that is uniform in* $t_0 \le t_n \le T$. *Then, for* C_τ *sufficiently small, the numerical method converges with order p. In particular, the numerical solution satisfies the error bound*

$$\|u_n - u(t_n)\| \le C \sum_{j=0}^{n-1} \tau_j^{p+1} \tag{2.21}$$

uniformly on $t_0 \le t_n \le T$. *The constant* C *is independent of the chosen step size sequence satisfying* (2.20).

Table 2   Stiff order conditions for exponential Rosenbrock methods applied to autonomous problems.

Number	Order condition	Order
1	$\sum_{i=1}^{s} b_i(z) = \varphi_1(z)$	1
2	$\sum_{j=1}^{i-1} a_{ij}(z) = c_i\varphi_1(c_iz),\quad 2 \le i \le s$	2
3	$\sum_{i=2}^{s} b_i(z)c_i^2 = 2\varphi_3(z)$	3
4	$\sum_{i=2}^{s} b_i(z)c_i^3 = 6\varphi_4(z)$	4

The well-known exponential Rosenbrock–Euler method (1.19) obviously satisfies condition 1 of Table 2, while condition 2 is void. Therefore, it is second-order convergent for problems satisfying our analytic framework. A possible error estimator for (1.19) is described in Caliari and Ostermann (2009).

Example 2.12. Hochbruck, Lubich, and Selhofer (1998) proposed the following class of exponential integrators

$$k_i = \varphi_1(-\gamma\tau A_n)\left(-A_n u_n + g_n(U_{ni}) - \tau A_n \sum_{j=1}^{i-1} \beta_{ij}k_j\right),$$

$$U_{ni} = u_n + \tau \sum_{j=1}^{i-1} \alpha_{ij}k_j, \quad i = 1,\ldots,s,$$

$$u_{n+1} = u_n + \tau \sum_{i=1}^{s} b_i k_i,$$

where $\gamma, \alpha_{ij}, \beta_{ij}, b_i$ are coefficients that determine the method.

Note that in contrast to general Rosenbrock methods, this method uses φ_1-function only. It thus cannot have order larger than two.

However, the method exp4 was designed such that the computation of the matrix functions is particularly efficient. To achieve this, it was not written as a three-stage exponential Rosenbrock-type method but as a seven-stage method, which uses only three function evaluations. The code comes with error and step size control and uses Krylov approximations for the approximation of the matrix functions.

The exp4 method of Hochbruck, Lubich, and Selhofer (1998) led to a revival of exponential integrators and initiated a lot of activities on the construction, implementation, analysis, and applications of such methods.

Examples of higher order exponential Rosenbrock-type methods and references to these methods can be found in Hochbruck, Ostermann, and Schweitzer (2009) and Hochbruck and Ostermann (2010). A comparative study of the performance of exponential, implicit, and explicit integrators for stiff systems of ordinary differential equations is given in Loffeld and Tokman (2013). Variants of exponential Runge–Kutta methods which are designed for efficient implementations have been proposed in Tokman (2006, 2011) and Tokman, Loffeld, and Tranquilli (2012). Higher order methods Rosenbrock methods are constructed and analyzed by Luan and Ostermann (2014b).

3   Approximation of the matrix exponential operator

This section deals with the approximation of products of a function of a matrix with a vector,

$$\mathbf{x} = \phi(-\tau\mathbf{A})\mathbf{b}, \qquad \mathbf{A} \in \mathbb{C}^{n,n}, \quad \mathbf{b} \in \mathbb{C}^n, \quad \|\mathbf{b}\| = 1, \quad \tau > 0 \qquad (3.1)$$

where $\phi : \mathbb{C} \to \mathbb{C}$ is analytic in a neighborhood of the spectrum of $-\tau\mathbf{A}$ or in some cases even in a neighborhood of the field of values of $-\tau\mathbf{A}$.

We refer to the excellent monograph of Higham (2008) for a detailed study of matrix functions. A survey on matrix functions with emphasis on the matrix exponential was given by Frommer and Simoncini (2008).

Here, we are interested in $\phi(z) = e^z$ or $\phi(z) = \varphi_k(z)$, $k = 1, 2, \ldots$, which arise in exponential integrators.

3.1 Arnoldi algorithm

Discretizations of partial differential operators lead to matrices $\mathbf{A}$ which are large and sparse. Since, in general, $\mathbf{A}^j$ is no longer sparse if j is large, the same holds for $\phi(-\tau\mathbf{A})$. In many applications, it is sufficient to compute $\phi(-\tau\mathbf{A})\mathbf{b}$ and this can be accomplished without computing $\phi(-\tau\mathbf{A})$ explicitly. Here, we are interested in approximations in certain Krylov subspaces.

There are several possible ways to motivate Krylov approximations to matrix functions. Here, we follow the derivation of Hochbruck and Lubich (1997). Let Γ be a contour surrounding $\mathcal{F}(-\tau\mathbf{A})$. Note that Cauchy's integral formula,

$$\phi(-\tau\mathbf{A})\mathbf{b} = \frac{1}{2\pi i}\int_{\Gamma}\phi(\lambda)(\lambda\mathbf{I}+\tau\mathbf{A})^{-1}\mathbf{b}\,d\lambda = \frac{1}{2\pi i}\int_{\Gamma}\phi(\lambda)\mathbf{x}(\lambda)\,d\lambda \quad (3.2)$$

is based on the solution of shifted linear systems

$$(\lambda\mathbf{I}+\tau\mathbf{A})\mathbf{x}(\lambda) = \mathbf{b}, \qquad \lambda \in \Gamma. \quad (3.3)$$

These linear systems can be approximated in a Krylov subspace defined as

$$\mathcal{K}_m(\mathbf{A},\mathbf{b}) = \mathrm{span}\{\mathbf{b},\mathbf{A}\mathbf{b},\ldots,\mathbf{A}^{m-1}\mathbf{b}\}. \quad (3.4)$$

We restrict ourselves to Arnoldi-based methods, which compute an orthonormal basis $\mathbf{V}_m \in \mathbb{C}^{n,m}$ of $\mathcal{K}_m(\mathbf{A},\mathbf{b})$. The derivation can be done for Lanczos-based methods in an analogous way. For Hermitian or skew-Hermitian matrices, the Lanczos and the Arnoldi algorithm are equivalent.

The Arnoldi recurrence reads

$$\mathbf{A}\mathbf{V}_m = \mathbf{V}_m\mathbf{H}_m + h_{m+1,m}\mathbf{v}_{m+1}\mathbf{e}_m^T, \qquad \mathbf{V}_m^H\mathbf{V}_m = \mathbf{I}_m,$$

where

$$\mathbf{V}_m^H\mathbf{A}\mathbf{V}_m = \mathbf{H}_m. \quad (3.5)$$

A simple implementation is given in the following algorithm.

Algorithm 3.1 (Arnoldi algorithm, see, *e.g.*, Saad (1992)).

Given $\mathbf{A} \in \mathbb{C}^{n,n}$, $\mathbf{b} \in \mathbb{C}^n$, and $\beta = \|\mathbf{b}\| > 0$
$\mathbf{v}_1 = \mathbf{b}/\beta$
for $m = 1, 2, \ldots$

- for $j = 1, \ldots, m$
    $h_{j,m} = \langle \mathbf{v}_j, \mathbf{A}\mathbf{v}_m \rangle$
- $\tilde{\mathbf{v}}_{m+1} = \mathbf{A}\mathbf{v}_m - \sum_{j=1}^m h_{j,m}\mathbf{v}_j$
- $h_{m+1,m} = \sqrt{\langle \tilde{\mathbf{v}}_{m+1}, \tilde{\mathbf{v}}_{m+1} \rangle}$

- $\mathbf{v}_{m+1} = \tilde{\mathbf{v}}_{m+1}/h_{m+1,m}$

Here, we will use the standard Euclidean inner product for the ease of presentation. However, the inner product within the Arnoldi algorithm should be chosen according to the application. In particular, for finite element disretizations, $\mathbf{V}_m$ should be orthonormal with respect to the discrete L^2 inner product, i.e., $\mathbf{V}_m^H \mathbf{M} \mathbf{V}_m = \mathbf{I}_m$. The algorithm and the results below can be adapted without any difficulty.

The following properties of the Arnoldi algorithm are easily verified.

Lemma 3.2. *Let $\mathbf{V}_m$ and $\mathbf{H}_m$ be the matrices from the Arnoldi algorithm applied to $\mathbf{A}$ and $\mathbf{b}$ with $\|\mathbf{b}\| = 1$. Then*

(a) $\mathcal{F}(\mathbf{H}_m) \subseteq \mathcal{F}(\mathbf{A})$,

(b) $p_{m-1}(\mathbf{A})\mathbf{b} = \mathbf{V}_m p_{m-1}(\mathbf{H}_m)\mathbf{e}_1$ *for all polynomials of degree at most $m - 1$.*

Proof. Exercise.   □

The definition of a Krylov subspace implies that

$$\mathcal{K}_m(\mathbf{A}, \mathbf{b}) = \mathcal{K}_m(\lambda \mathbf{I} + \tau \mathbf{A}, \mathbf{b}) \qquad \text{for all} \qquad \lambda \in \mathbb{C}.$$

Moreover, the relation

$$(\lambda \mathbf{I} + \tau \mathbf{A})\mathbf{V}_m = \mathbf{V}_m(\lambda \mathbf{I} + \tau \mathbf{H}_m) + \tau h_{m+1,m}\mathbf{v}_{m+1}\mathbf{e}_m^T$$

holds.

The Galerkin approximation $\mathbf{x}_m(\lambda) \in \mathcal{K}_m(\mathbf{A}, \mathbf{b})$ for the solution $\mathbf{x}(\lambda)$ of (3.3) is defined by the condition that the residual

$$\mathbf{r}_m(\lambda) = \mathbf{b} - (\lambda \mathbf{I} + \tau \mathbf{A})\mathbf{x}_m(\lambda) \tag{3.6}$$

is orthogonal to $\mathcal{K}_m(\mathbf{A}, \mathbf{b})$. Writing $\mathbf{x}_m(\lambda) = \mathbf{V}_m \mathbf{y}_m(\lambda)$, this is equivalent to

$$0 = \mathbf{V}_m^H \mathbf{r}_m(\lambda) = \mathbf{e}_1 - (\lambda \mathbf{I} + \tau \mathbf{H}_m)\mathbf{y}_m(\lambda).$$

If we choose a curve Γ which includes the field of values $\mathcal{F}(-\tau \mathbf{A})$ in its interior, and by Lemma 3.2 also $\mathcal{F}(-\tau \mathbf{H}_m)$, then Γ does not contain any eigenvalue of $-\tau \mathbf{H}_m$. Hence, $\lambda \mathbf{I} + \tau \mathbf{H}_m$ is nonsingular and

$$\mathbf{x}_m(\lambda) = \mathbf{V}_m(\lambda \mathbf{I} + \tau \mathbf{H}_m)^{-1}\mathbf{e}_1. \tag{3.7}$$

This is the approximation of the full orthogonalization method (FOM) by Saad (1981).

An approximation to $\phi(-\tau\mathbf{A})\mathbf{b}$ is obtained by approximating $\mathbf{x}_m(\lambda)$ within the Cauchy integral formula by its FOM iterate:

$$
\begin{aligned}
\phi(-\tau\mathbf{A})\mathbf{b} &\approx \frac{1}{2\pi i}\int_\Gamma \phi(\lambda)\mathbf{x}_m(\lambda)d\lambda \\
&= \frac{1}{2\pi i}\int_\Gamma \phi(\lambda)\mathbf{V}_m(\lambda\mathbf{I}+\tau\mathbf{H}_m)^{-1}\mathbf{e}_1 d\lambda \qquad (3.8) \\
&= \mathbf{V}_m\frac{1}{2\pi i}\int_\Gamma \phi(\lambda)(\lambda\mathbf{I}+\tau\mathbf{H}_m)^{-1}d\lambda\,\mathbf{e}_1 \\
&= \mathbf{V}_m\phi(-\tau\mathbf{H}_m)\mathbf{e}_1.
\end{aligned}
$$

The last identity is again a Cauchy integral formula, but this time for the tiny matrix $\mathbf{H}_m$. For such a matrix, $\phi(-\tau\mathbf{H}_m)$ can be computed or approximated explicitly by diagonalization of $\mathbf{H}_m$ or by Padé approximation.

Note that we used the solution of the linear systems only for the purpose of deriving the approximation $\mathbf{x}_m$. In practice, we do not have to solve any of these linear systems but instead we just compute $\phi(-\tau\mathbf{H}_m)$. However, it is worth mentioning that one could also choose certain nodes θ_k on the curve Γ and construct a quadrature formula, which then requires the solution of a small number of linear systems to approximate $\mathbf{x}_m(\theta_k)$, see, *e.g.*, Trefethen, Weideman, and Schmelzer (2006) and Hale, Higham, and Trefethen (2008).

3.2   Stopping criteria

The above derivation via the Cauchy integral formula also motivates a stopping criterion for approximations to matrix functions, cf. Hochbruck and Lubich (1997), section 6.3. For linear systems

$$(\lambda\mathbf{I}+\tau\mathbf{A})\mathbf{x}(\lambda) = \mathbf{b}$$

it is usually based on the residual $\mathbf{r}_m(\lambda)$ instead of the error

$$\epsilon_m(\lambda) = \mathbf{x}_m(\lambda) - \mathbf{x}(\lambda).$$

The residuals for the linear systems satisfy

$$\mathbf{r}_m(\lambda) = \tau h_{m+1,m}\big(\mathbf{e}_m^T(\lambda\mathbf{I}+\tau\mathbf{H}_m)^{-1}\mathbf{e}_1\big)\mathbf{v}_{m+1}.$$

Using Cauchy's integral formula, the error of the mth Krylov approximation to $\phi(-\tau\mathbf{A})$ can be written as

$$\epsilon_m = \mathbf{V}_m\phi(-\tau\mathbf{H}_m)\mathbf{e}_1 - \phi(-\tau\mathbf{A})\mathbf{b} = \frac{1}{2\pi i}\int_\Gamma \phi(\lambda)\epsilon_m(\lambda)d\lambda.$$

Replacing the error $\epsilon_m(\lambda)$ by $\mathbf{r}_m(\lambda)$, we get the *generalized residual*

$$\mathbf{r}_m = \frac{1}{2\pi i} \int_\Gamma \phi(\lambda)\mathbf{r}_m(\lambda)d\lambda = \tau h_{m+1,m}\left(\mathbf{e}_m^T\phi(-\tau\mathbf{H}_m)\mathbf{e}_1\right)\mathbf{v}_{m+1},$$

which can be computed at no additional cost. The same stopping criterion was proposed earlier by Saad (1992) with a different motivation and in Lubich (2008, p. 94), a refined version can be found for the case of $\phi(z) = e^z$ and skew-Hermitian matrices $\mathbf{A}$. A more detailed discussion on stopping criteria can be found in Botchev, Grimm, and Hochbruck (2013).

In Hochbruck and Lubich (1997) it was shown how to derive error bounds for the approximation of the matrix exponential based on Cauchy's integral formula by using Faber polynomials for different sets in the complex plane. For the special case of skew Hermitian matrices, error bounds and stopping criteria are given in Lubich (2008). Here, the analysis is based on Chebyshev series.

3.3 Software

For the polynomial approximation of the matrix exponential, there is excellent MATLAB software available. Here is a list of packages, which are freely available.

- expmv: MATLAB code by Al-Mohy and Higham (2011). It is based on a Taylor expansion and includes applications to exponential integrators.

- expmvp and phipm: MATLAB code by Niesen and Wright (2012). It contains an implementation of Krylov subspace methods for approximating φ-functions

- expokit: MATLAB and Fortran codes by Sidje (1998). This code approximates $e^{\tau\mathbf{A}}\mathbf{b}$ and uses a time-stepping procedure, if the number of iterations exceeds a given limit (default value is $m_{\max} = 30$).

Acknowledgement

The author would like to thank Volker Grimm and Tanja Göckler for their help in preparing the lectures for the Gene Golub SIAM Summer School in Shanghai 2013 and for their careful reading of this paper.

References

A. H. Al-Mohy and N. J. Higham. Computing the action of the matrix exponential, with an application to exponential integrators. *SIAM J. Sci. Comp.*, 33(2): 488–511, 2011. URL `http://link.aip.org/link/?SCE/33/488/1`.

H. Berland, B. Skaflestad, and W. M. Wright. EXPINT—A MATLAB package for exponential integrators. *ACM Trans. Math. Softw.*, 33(1), Mar. 2007. URL `http://doi.acm.org/10.1145/1206040.1206044`.

M. A. Botchev, V. Grimm, and M. Hochbruck. Residual, Restarting, and Richardson Iteration for the Matrix Exponential. *SIAM J. Sci. Comput.*, 35(3): A1376–A1397, 2013. URL `http://dx.doi.org/10.1137/110820191`.

M. Caliari and A. Ostermann. Implementation of exponential Rosenbrock-type integrators. *Appl. Numer. Math.*, 59(3-4): 568–581, 2009. URL `http://www.sciencedirect.com/science/article/B6TYD-4S3WWWF-2/2/2812d6299c63c61465401cd7e7085815`.

K.-J. Engel and R. Nagel. *One-parameter semigroups for linear evolution equations*, volume 194 of *Graduate Texts in Mathematics*. Springer-Verlag, New York, 2000. With contributions by S. Brendle, M. Campiti, T. Hahn, G. Metafune, G. Nickel, D. Pallara, C. Perazzoli, A. Rhandi, S. Romanelli and R. Schnaubelt.

K.-J. Engel and R. Nagel. *A short course on operator semigroups*. Universitext. Springer, New York, 2006.

A. Frommer and V. Simoncini. Matrix functions. In *Model order reduction: theory, research aspects and applications*, volume 13 of *Math. Ind.*, pages 275–303. Springer, Berlin, 2008. URL `http://dx.doi.org/10.1007/978-3-540-78841-6_13`.

N. Hale, N. J. Higham, and L. N. Trefethen. Computing $\mathbf{A}^\alpha$, $\log(\mathbf{A})$, and related matrix functions by contour integrals. *SIAM J. Numer. Anal.*, 46 (5): 2505–2523, 2008. URL `http://dx.doi.org/10.1137/070700607`.

N. J. Higham. *Functions of Matrices: Theory and Computation*. SIAM, Philadelphia, 2008.

M. Hochbruck and C. Lubich. On Krylov subspace approximations to the matrix exponential operator. *SIAM J. Numer. Anal.*, 34(5): 1911–1925, 1997. URL `http://dx.doi.org/10.1137/S0036142995280572`.

M. Hochbruck, C. Lubich, and H. Selhofer. Exponential integrators for large systems of differential equations. *SIAM J. Sci. Comput.*, 19(5): 1552–1574, 1998. URL `http://link.aip.org/link/?SCE/19/1552/1`.

M. Hochbruck and A. Ostermann. Explicit exponential Runge–Kutta methods for semilinear parabolic problems. *SIAM J. Numer. Anal.*, 43(3): 1069–1090, 2005. URL `http://link.aip.org/link/?SNA/43/1069/1`.

M. Hochbruck and A. Ostermann. Exponential integrators. *Acta Numerica*, 19: 209–286, 2010. URL http://journals.cambridge.org/action/ displayAbstract?fromPage=online&aid=7701740&fulltextType=RA& fileId=S0962492910000048.

M. Hochbruck, A. Ostermann, and J. Schweitzer. Exponential Rosenbrock-type methods. *SIAM J. Numer. Anal.*, 47(1): 786–803, 2009. URL http://link.aip.org/link/?SNA/47/786/1.

J. Loffeld and M. Tokman. Comparative performance of exponential, implicit, and explicit integrators for stiff systems of ODEs. *J. Comput. Appl. Math.*, 241: 45–67, 2013. URL http://dx.doi.org/10.1016/j.cam.2012.09.038.

V. T. Luan and A. Ostermann. Explicit exponential Runge–Kutta methods of high order for parabolic problems. *J. Comput. Appl. Math.*, 256: 168–179, 2014a. URL http://dx.doi.org/10.1016/j.cam.2013.07.027.

V. T. Luan and A. Ostermann. Exponential Rosenbrock methods of order five — construction, analysis and numerical comparisons. *J. Comput. Appl. Math.*, 255: 417–431, 2014b. URL http://dx.doi.org/10.1016/j.cam. 2013.04.041.

C. Lubich. *From Quantum to Classical Molecular Dynamics: Reduced Models and Numerical Analysis*. Zurich Lectures in Advanced Mathematics. European Mathematical Society (EMS), 2008.

J. Niesen and W. M. Wright. Algorithm 919: A Krylov subspace algorithm for evaluating the φ-functions appearing in exponential integrators. *ACM Trans. Math. Softw.*, 38(3): 22:1–22:19, Apr. 2012. URL http://doi.acm. org/10.1145/2168773.2168781.

A. Pazy. *Semigroups of Linear Operators and Applications to Partial Differential Equations*. Number v. 44 in Applied Mathematical Sciences. Springer, 1992. URL http://books.google.de/books?id=sIAyOgM4R3kC.

D. A. Pope. An exponential method of numerical integration of ordinary differential equations. *Comm. ACM*, 6: 491–493, 1963.

Y. Saad. Krylov subspace methods for solving large unsymmetric linear systems. *Math. Comp.*, 37(155): 105–126, 1981. URL http://dx.doi.org/10. 2307/2007504.

Y. Saad. Analysis of some Krylov subspace approximations to the matrix exponential operator. *SIAM J. Numer. Anal.*, 29(1): 209–228, 1992. URL http://dx.doi.org/10.1137/0729014.

R. B. Sidje. EXPOKIT: a software package for computing matrix exponentials. *ACM Trans. Math. Softw.*, 24(1): 130–156, Mar. 1998. URL http://doi. acm.org/10.1145/285861.285868.

M. Tokman. Efficient integration of large stiff systems of ODEs with exponential propagation iterative (EPI) methods. *J. Comput. Phys.*, 213(2): 748–776, 2006. URL http://dx.doi.org/10.1016/j.jcp.2005.08.032.

M. Tokman. A new class of exponential propagation iterative methods of Runge-Kutta type (EPIRK). *J. Comput. Phys.*, 230(24): 8762–8778, 2011. URL http://dx.doi.org/10.1016/j.jcp.2011.08.023.

M. Tokman, J. Loffeld, and P. Tranquilli. New adaptive exponential propagation iterative methods of Runge-Kutta type. *SIAM J. Sci. Comput.*, 34(5): A2650–A2669, 2012. URL http://dx.doi.org/10.1137/110849961.

L. N. Trefethen, J. A. C. Weideman, and T. Schmelzer. Talbot quadratures and rational approximations. *BIT*, 46(3): 653–670, 2006. URL http://dx.doi.org/10.1007/s10543-006-0077-9.

Matrix Equations and Model Reduction

Peter Benner*   Tobias Breiten†   Lihong Feng‡

Abstract

We review model order reduction methods for linear time invariant systems and the numerical solution of related matrix equations. The basic ideas of the methods, such as Padé approximation, moment matching, rational interpolation, modal truncation, balanced truncation and related methods, are presented. The numerical algorithms used for implementing the methods are discussed. For balanced truncation and the balancing related methods, Lyapunov equations or Riccati equations need to be solved. Algorithms for solving these matrix equations are introduced.

1 Introduction

Model order reduction (MOR) is a computational technique for reducing the complexity of simulating large-scale complex systems, so that the input-output relations can be reproduced in acceptable time and with ignorable error. In today's real-life applications, large-scale complex systems can be time-varying, nonlinear, parametric, or stochastic, which pose big challenges for model order reduction. Although model order reduction techniques have been developed for these systems, and proved to be promising in various applications, for this introductory course we focus on model order reduction methods for linear time invariant (LTI) systems. MOR methods for more sophisticated problem formulations can be derived by using the same basic ideas as outlined here. Hence, we consider systems in the following form:

$$\dot{x}(t) = Ax(t) + Bu(t), \quad x(0) = x_0 \in \mathbb{R}^n,$$
$$y(t) = Cx(t) + Du(t), \tag{1.1}$$

with $A \in \mathbb{R}^{n \times n}$, $B \in \mathbb{R}^{n \times m}$, $C \in \mathbb{R}^{p \times n}$ and $D \in \mathbb{R}^{p \times m}$. Here, $x(t) \in \mathbb{R}^n$ is the state of the system, $u(t) \in \mathbb{R}^m$ is the input, and $y(t) \in \mathbb{R}^p$ is the output. When $m = p = 1$, the system is called single-input single-output

Max-Planck Institute for Dynamics of Complex Technical Systems, Sandtorstr.1, 39106 Magdeburg, Germany. E-mail: *benner@mpi-magdeburg.mpg.de, †breiten@mpi-magdeburg.mpg.de, ‡feng@mpi-magdeburg.mpg.de.

(SISO) system; otherwise if $m, p > 1$, it is called a multiple-input and multiple-output (MIMO) system.

The basic idea of model order reduction is based on projection. Assume that the trajectory of x in (1.1) is (approximately) contained in a low-dimensional subspace $\mathcal{V}$, and $\mathcal{W}^{\perp}$ is a complementary subspace of $\mathcal{V}$, i.e. $\mathcal{V} \oplus \mathcal{W}^{\perp} = \mathbb{R}^n$, $\mathcal{V} \cap \mathcal{W}^{\perp} = \{0\}$. Let $\mathcal{W}$ be the orthogonal complement of $\mathcal{W}^{\perp}$, and let the columns of the matrix $V \in \mathbb{R}^{n \times q}$ form the basis of $\mathcal{V}$, let the columns of $W \in \mathbb{R}^{n \times q}$ be the basis of the subspace $\mathcal{W}$, such that $W^T V = I$, then $V W^T$ is a projector, which projects x onto $\mathcal{V}$ along $\mathcal{W}^{\perp}$. The reduced-order model is obtained by approximating the state x by its projection $x \approx V W^T x$,

$$
\begin{aligned}
\dot{\hat{x}}(t) &= \hat{A}\hat{x}(t) + \hat{B}u(t), \\
\hat{y}(t) &= \hat{C}\hat{x}(t) + \hat{D}u(t),
\end{aligned}
\tag{1.2}
$$

where $\hat{x}(t) = W^T x(t) \in \mathbb{R}^q$, $\hat{A} = W^T A V \in \mathbb{R}^{q \times q}$, $\hat{B} = W^T B \in \mathbb{R}^{q \times m}$, $\hat{C} = CV \in \mathbb{R}^{p \times m}$, and $\hat{D} = D \in \mathbb{R}^{p \times m}$. The above process of getting the reduced model is in fact a Petrov-Galerkin projection. This can be seen by replacing x with the approximation $x \approx V\hat{x} =: \tilde{x}$ and then forcing the residual $r = \dot{\tilde{x}} - A\tilde{x} - Bu$ to be zero in the test subspace $\mathcal{W}$, i.e. $W^T r = 0$, so that the first equation in (1.2) is derived. The second equation follows directly by replacing x with its approximation $\tilde{x} = V\hat{x}$. When $W = V$, it reduces to a Galerkin projection.

The goals of model order reduction method include

- The output of the large-scale system should be approximated by a reduced model that can be evaluated significantly faster.

- The reduced model should be automatically generated.

- There should be a computable error bound/estimate for the reduced model.

- Physical properties of the original system, such as stability, minimum phase, and/or passivity should be preserved during the MOR process.

The MOR methods discussed here are based on concepts from (numerical) linear algebra and systems and control theory, where matrix decompositions, Krylov subspaces, iterative solvers, matrix equations play important roles. The outline of this chapter is as follows. In the next section, the mathematical basics are summarized. In sections 3–6, basic model reduction methods for LTI systems are presented. Numerical algorithms for solving matrix equations are discussed in section 7. Model reduction related software is introduced in section 8. Conclusions are given in the end.

2   Mathematical basics

In this section we will provide basic facts from various mathematical disciplines used in the following. We refrain from providing references, all this material can be found in pertinent textbooks and is also collected in the seminal textbook on MOR by Antoulas, [1].

The singular value decomposition

One essential tool from (numerical) linear algebra for data compression and dimension reduction is the singular value decomposition (SVD) of a matrix. The SVD exists for any matrix as the following theorem shows.

Theorem 1. *Let $A \in \mathbb{R}^{m \times n}$, then there exist orthogonal $U \in \mathbb{R}^{m \times m}$ and $V \in \mathbb{R}^{n \times n}$, such that*

$$A = U \Sigma V^T, \quad \Sigma = \begin{cases} \begin{bmatrix} \Sigma_1 \\ 0 \end{bmatrix}, & m \geq n, \\ \begin{bmatrix} \Sigma_1 & 0 \end{bmatrix}, & m \leq n, \end{cases}$$

where

$$\Sigma_1 = \begin{bmatrix} \sigma_1 & & \\ & \ddots & \\ & & \sigma_{\min(m,n)} \end{bmatrix}$$

and $\sigma_1 \geq \cdots \geq \sigma_s > \sigma_{s+1} = \cdots = \sigma_{\min(m,n)} = 0$ for $s = \operatorname{rank}(A)$.

The singular value decomposition of matrices is the core of the balanced truncation MOR method. It is also used in many other model reduction methods to assist the derivation of the reduced model.

The Laplace transform

Definition 1. *The* Laplace transform *of a time domain function $f \in L_{1,\text{loc}}$ (f is locally integrable, i.e., $\int_K |f(t)| dt < \infty$, $\forall$ compact subsets K of $\operatorname{dom}(f)$) with $\operatorname{dom}(f) = \mathbb{R}_0^+$ is*

$$\mathcal{L} : f(t) \mapsto F(s) := \mathcal{L}\{f(t)\}(s) := \int_0^\infty e^{-st} f(t)\, dt, \quad s \in \mathbb{C}.$$

F is a function in the (Laplace or) frequency domain.

For frequency domain evaluations ("frequency response analysis"), one takes $\operatorname{Re}(s) = 0$ and $\operatorname{Im}(s) \geq 0$. Then $\omega := \operatorname{Im}(s)$ takes the role of a frequency (in [rad/s], i.e., $\omega = 2\pi v$ with v measured in [Hz]).

Lemma 1. *Applying the Laplace transform to the derivative of $f(t)$ results in $sF(s)$,*

$$\mathcal{L}\{\dot{f}(t)\}(s) = sF(s).$$

For ease of notation, in the following we will use lower-case letters for both the function and its Laplace transform.

Linear systems in frequency domain

Applying the Laplace transform $(x(t) \mapsto x(s),\ \dot{x}(t) \mapsto sx(s))$ to the linear system in (1.1) with $x(0) = 0$ yields

$$sx(s) = Ax(s) + Bu(s), \quad y(s) = Cx(s) + Du(s).$$

We get the input-output relation in frequency domain,

$$y(s) = \Big(\underbrace{C(sI - A)^{-1}B + D}_{=:G(s)} \Big)u(s),$$

where $G(s)$ is defined as the transfer function of (1.1).

In systems and control theory, the error bound of the reduced-order model is established through the transfer function, e.g.

$$||y - \hat{y}||_2 \le ||G(s) - \hat{G}(s)||_\infty ||u||_2,$$

where the 2-norm stands for the $\mathcal{L}_2$ (or $\mathcal{H}_2$) norm in the frequency domain, or the L_2 norm in the time domain. $||\cdot||_\infty$ is the $\mathcal{H}_\infty$ norm of a matrix-valued function (see the analysis in the subsection "System norms"). $\hat{G}(s) = \hat{C}(sI - \hat{A})^{-1}B + \hat{D}$ is the transfer function of the reduced-order model. The details of deriving the error bound are discussed at the end of this section.

Properties of linear systems

Definition 2. *A linear system*

$$\dot{x}(t) = Ax(t) + Bu(t), \quad y(t) = Cx(t) + Du(t)$$

is *stable if its transfer function $G(s)$ has all its poles in the left half plane and it is* asymptotically *(or Lyapunov, or exponentially) stable if all poles are in the open left half plane $\mathbb{C}^- := \{z \in \mathbb{C} \mid \operatorname{Re}(z) < 0\}$.*

Lemma 2. *A sufficient condition for asymptotic stability is that A is asymptotically stable (or Hurwitz), i.e., the spectrum of A, denoted by $\Lambda(A)$, satisfies $\Lambda(A) \subset \mathbb{C}^-$.*

Note that by abuse of notation, often "stable system" is used for asymptotically stable systems. In what follows, we need to define the concepts of controllability and observability.

Definition 3. *Given a linear system* (A, B, C, D). *A state* $x_* \in \mathbb{R}^n$ *is* controllable to zero *if for all initial values* $x_0 \in \mathbb{R}^n$, *there exist an input function* $u_*(t)$ *and a time* $t_* < \infty$, *such that the solution of the linear dynamical system vanishes at time* t_*, *i.e.,* $\Phi(u_*; x_*; t_*) = 0$. *(Here,* $\Phi(u_*; x_*; t_*) = 0$ *denotes the solution of* (1.1) *for a given input function* u_* *at time* $t = t_*$.)

The controllable subspace X^{contr} *of the system is the set of all controllable states. The system is* (completely) controllable *if* $X^{\text{contr}} = \mathbb{R}^n$.

Definition 4. *Given a linear system* (A, B, C, D). *A state* $x_* \in \mathbb{R}^n$ *is* unobservable *if* $y(t) = 0$ *for* $x_0 = x_*$ *and all* $t \geq 0$, *i.e., if* x_* *is indistinguishable from the zero state for all* $t \geq 0$. *The* unobservable subspace X^{unobs} *is the set of all unobservable states of the system. The system is* (completely) observable *if* $X^{\text{unobs}} = \{0\}$.

Controllability and observability, characterized by the *controllability matrix* $K(A, B) = [B, AB, A^2B, \ldots, A^{n-1}B] \in \mathbb{R}^{n \times nm}$, and the *observability matrix*

$$\mathcal{O}(A, C) = \begin{bmatrix} C \\ CA \\ CA^2 \\ \vdots \\ CA^{n-1} \end{bmatrix} \in \mathbb{R}^{np \times n},$$

are two important properties of the system, based on which the standard balanced truncation method and balancing related MOR methods are developed.

Lemma 3. *The LTI system* (1.1) *is controllable if and only if* $K(A, B)$ *has full rank* n. *Analogously, it is observable if and only if* $\mathcal{O}(A, C)$ *has full rank* n.

The controllability and observability of (1.1) in case of stable LTI systems can also be examined through the infinite Gramians P and Q of the system. The *controllability Gramian* P and the *observability Gramian* Q of (1.1) can be defined via

$$\begin{aligned} P &= \int_0^\infty e^{At} B B^T e^{A^T t} dt, \\ Q &= \int_0^\infty e^{A^T t} C^T C e^{At} dt. \end{aligned} \tag{2.1}$$

Lemma 4. *The LTI system* (1.1) *is controllable if and only if P is positive definite. It is observable if and only if Q is positive definite.*

Please refer to the pertinent literature, e.g., [1], for more discussion on controllability and observability, and other properties of linear systems, such as stabilizability, detectability etc.

Realizations of linear systems

Definition 5. *For a linear time-invariant system*

$$\Sigma : \begin{cases} \dot{x}(t) = Ax(t) + Bu(t), \\ y(t) = Cx(t) + Du(t) \end{cases}$$

with transfer function $G(s) = C(sI - A)^{-1}B + D$, *the quadruple* $(A, B, C, D) \in \mathbb{R}^{n \times n} \times \mathbb{R}^{n \times m} \times \mathbb{R}^{p \times n} \times \mathbb{R}^{p \times m}$ *is called a realization of* Σ.

It can be easily verified that the transfer function is invariant under state-space transformations,

$$\mathcal{T} : \begin{cases} x & \rightarrow & Tx, \\ (A, B, C, D) & \rightarrow & (TAT^{-1}, TB, CT^{-1}, D). \end{cases}$$

The transfer function is also invariant under addition of uncontrollable or unobservable states as below:

$$\frac{d}{dt}\begin{bmatrix} x \\ x_1 \end{bmatrix} = \begin{bmatrix} A & 0 \\ 0 & A_1 \end{bmatrix} \begin{bmatrix} x \\ x_1 \end{bmatrix} + \begin{bmatrix} B \\ B_1 \end{bmatrix} u(t),$$

$$y(t) = \begin{bmatrix} C & 0 \end{bmatrix} \begin{bmatrix} x \\ x_1 \end{bmatrix} + Du(t),$$

or

$$\frac{d}{dt}\begin{bmatrix} x \\ x_2 \end{bmatrix} = \begin{bmatrix} A & 0 \\ 0 & A_2 \end{bmatrix} \begin{bmatrix} x \\ x_2 \end{bmatrix} + \begin{bmatrix} B \\ 0 \end{bmatrix} u(t),$$

$$y(t) = \begin{bmatrix} C & C_2 \end{bmatrix} \begin{bmatrix} x \\ x_2 \end{bmatrix} + Du(t),$$

for arbitrary $A_j \in \mathbb{R}^{n_j \times n_j}$, $j = 1, 2$, $B_1 \in \mathbb{R}^{n_1 \times m}$, $C_2 \in \mathbb{R}^{p \times n_2}$ and any $n_1, n_2 \in \mathbb{N}$. Hence, the following four quadruples

$$(A, B, C, D), \qquad \left(\begin{pmatrix} A & 0 \\ 0 & A_1 \end{pmatrix}, \begin{pmatrix} B \\ B_1 \end{pmatrix}, (C\ 0), D \right),$$

$$(TAT^{-1}, TB, CT^{-1}, D), \qquad \left(\begin{pmatrix} A & 0 \\ 0 & A_2 \end{pmatrix}, \begin{pmatrix} B \\ 0 \end{pmatrix}, (C\ C_2), D \right)$$

are all realizations of Σ. Therefore, realizations of an LTI system are not unique, and not even their orders need to coincide. The latter problem can be fixed as there is a lower bound on the possible order of a realization.

Definition 6. *The* McMillan degree *of* Σ *is the unique minimal number* $\hat{n} \geq 0$ *of states necessary to describe the input-output behavior completely. A* minimal realization *is a realization* $(\hat{A}, \hat{B}, \hat{C}, \hat{D})$ *of* Σ *with order* $\hat{n}$.

Theorem 2. *A realization* (A, B, C, D) *of a linear system is minimal if and only if it is controllable and observable.*

Balanced realizations

Definition 7. *A realization* (A, B, C, D) *of a stable linear system* Σ *is* balanced *if its controllability/observability Gramians* P, Q *satisfy*

$$P = Q = \mathrm{diag}\left\{\sigma_1, \ldots, \sigma_n\right\} \quad (\textit{w.l.o.g.} \ \sigma_j \geq \sigma_{j+1}, \ j = 1, \ldots, n-1).$$

Notice that $\sigma_1, \ldots, \sigma_n \geq 0$ as $P, Q \geq 0$ by definition, and $\sigma_1, \ldots, \sigma_n > 0$ in case of minimality. In general, even for unbalanced systems, the so-called *Hankel singular values* σ_i^{HSV} can be computed by means of the Gramians P and Q. We have $\sigma_i^{\mathrm{HSV}} = \Lambda_i(PQ)^{\frac{1}{2}}$, i.e., the Hankel singular values are given as the positive square roots of the eigenvalues of the product of the Gramians P and Q. For more information on the precise definition of the Hankel singular values and their relation to the Hankel operator of the system, we refer to, e.g., [1]. The following theorem shows how to obtain a balanced realization. Assume A is Hurwitz, i.e., $\Lambda(A) \subset \mathbb{C}^-$. Then:

Theorem 3. *Given a stable minimal linear system* $\Sigma : (A, B, C, D)$, *a balanced realization is obtained by the state-space transformation with*

$$T_b := \Sigma^{-\frac{1}{2}} V^T R,$$

where $P = S^T S$, $Q = R^T R$ *(e.g., Cholesky decompositions) and* $SR^T = U\Sigma V^T$ *is the SVD of* SR^T.

In order to make use of balancing transformations, one has to compute the Gramians of the system. For this, their following characterization turns out to be useful.

Theorem 4. *The controllability/observability Gramians* P/Q *satisfy the Lyapunov equations*

$$AP + PA^T + BB^T = 0, \quad A^T Q + QA + C^T C = 0. \tag{2.2}$$

In the following, only the case for the controllability Gramian is proved for instructional purposes; the proof for the observability Gramian is analogous.

Proof. From the definition of P in (2.1),

$$AP + PA^T + BB^T$$

$$= A \int_0^\infty e^{At} BB^T e^{A^T t} dt + \int_0^\infty e^{At} BB^T e^{A^T t} dt \, A^T + BB^T$$

$$= \int_0^\infty \underbrace{A e^{At} BB^T e^{A^T t} + e^{At} BB^T e^{A^T t} A^T}_{= \frac{d}{dt} e^{At} BB^T e^{A^T t}} dt + BB^T$$

$$= \underbrace{\lim_{t \to \infty} e^{At} BB^T e^{A^T t}}_{=0} - \underbrace{e^{A \cdot 0}}_{= I_n} BB^T \underbrace{e^{A^T \cdot 0}}_{= I_n} + BB^T$$

$$= 0. \qquad\qquad \square$$

Theorem 5. *The Hankel singular values (HSVs) of a stable minimal linear system are system invariants, i.e., they are unaltered by state-space transformations.*

Proof. In balanced coordinates, the HSVs are $\Lambda(PQ)^{\frac{1}{2}}$. Now let

$$(\hat{A}, \hat{B}, \hat{C}, D) = (TAT^{-1}, TB, CT^{-1}, D)$$

be any transformed realization with associated controllability Lyapunov equation

$$0 = \hat{A}\hat{P} + \hat{P}\hat{A}^T + \hat{B}\hat{B}^T = TAT^{-1}\hat{P} + \hat{P}T^{-T}A^T T^T + TBB^T T^T.$$

This is equivalent to

$$0 = A(T^{-1}\hat{P}T^{-T}) + (T^{-1}\hat{P}T^{-T})A^T + BB^T.$$

The uniqueness of the solution of the Lyapunov equation[①] implies that $\hat{P} = TPT^T$ and, analogously, $\hat{Q} = T^{-T}QT^{-1}$. Therefore,

$$\hat{P}\hat{Q} = TPQT^{-1},$$

showing that $\Lambda(\hat{P}\hat{Q}) = \Lambda(PQ) = \{\sigma_1^2, \ldots, \sigma_n^2\}$. $\qquad\qquad \square$

For non-minimal systems, the Gramians can also be transformed into diagonal matrices with leading $\hat{n} \times \hat{n}$ submatrices equal to $\mathrm{diag}(\sigma_1, \ldots, \sigma_{\hat{n}})$ and

$$\hat{P}\hat{Q} = \mathrm{diag}(\sigma_1^2, \ldots, \sigma_{\hat{n}}^2, 0, \ldots, 0),$$

see [23, 31].

[①] Observe that the Lyapunov equation defines a regular linear system of equations due to the stability assumption, see Section 7.

System norms

Definition 8. *The $L_2^n(-\infty, +\infty)$ space is the vector-valued function space $f : \mathbb{R} \mapsto \mathbb{R}^n$, endowed with the norm*

$$\|f\|_{L_2^n} = \left(\int_{-\infty}^{\infty} \|f(t)\|^2 dt \right)^{1/2}.$$

Here and below, $\| \cdot \|$ denotes the Euclidean vector or spectral matrix norm.

Definition 9. *The frequency domain $\mathcal{L}_2(\jmath\mathbb{R})$ space is the matrix-valued function space $F : \mathbb{C} \mapsto \mathbb{C}^{p \times m}$, endowed with the norm*

$$\|F\|_{\mathcal{L}_2} = \left(\frac{1}{2\pi} \int_{-\infty}^{\infty} \|F(\jmath\omega)\|^2 d\omega \right)^{1/2},$$

where $\jmath = \sqrt{-1}$ is the imaginary unit.

The maximum modulus theorem [24] will be used in this subsection.

Theorem 6. *Let $f(z) : \mathbb{C}^n \mapsto \mathbb{C}$ be a regular analytic, or holomorphic, function of n complex variables $z = (z_1, \ldots, z_n), n \geq 1$, defined on an (open) domain $\mathbb{D}$ of the complex space $\mathbb{C}^n$, which is not a constant, $f(z) \neq const.$ Let*

$$\max_f = \sup\{|f(z)| : z \in \mathbb{D}\}.$$

Then $\max_f$ can only be attained on the boundary of $\mathbb{D}$.

Consider the transfer function

$$G(s) = C(sI - A)^{-1}B + D$$

and input functions $u \in \mathcal{L}_2(\jmath\mathbb{R})$, with the $\mathcal{L}_2$-norm

$$\|u\|_{\mathcal{L}_2}^2 := \frac{1}{2\pi} \int_{-\infty}^{\infty} u(\jmath\omega)^H u(\jmath\omega)\, d\omega.$$

Assume A is (asymptotically) stable: $\Lambda(A) \subset \mathbb{C}^- := \{z \in \mathbb{C} : \mathrm{Re}\,(z) < 0\}$. Then G is analytic in $\mathbb{C}^+ \cup \jmath\mathbb{R}$, and following the maximum modulus theorem, $G(s)$ is bounded: $\|G(s)\| \leq M < \infty\ \forall s \in \mathbb{C}^+ \cup \jmath\mathbb{R}$. Thus we have

$$\int_{-\infty}^{\infty} y(\jmath\omega)^H y(\jmath\omega)\, d\omega = \int_{-\infty}^{\infty} u(\jmath\omega)^H G(\jmath\omega)^H G(\jmath\omega) u(\jmath\omega)\, d\omega$$

$$= \int_{-\infty}^{\infty} \|G(\jmath\omega)u(\jmath\omega)\|^2\, d\omega$$

$$\leq \int_{-\infty}^{\infty} M^2 \|u(\jmath\omega)\|^2 d\omega$$

$$= M^2 \int_{-\infty}^{\infty} u(\jmath\omega)^H u(\jmath\omega)\, d\omega\ < \infty,$$

hence $y = Gu \in \mathcal{L}_2(\jmath\mathbb{R})$.

Consequently, the $\mathcal{L}_2$-induced operator norm

$$||G||_\infty := \sup_{||u||_2 \neq 0} \frac{||Gu||_{\mathcal{L}_2}}{||u||_{\mathcal{L}_2}} \qquad (2.3)$$

is well defined. It can be further proved that

$$||G||_\infty = \sup_{w \in \mathbb{R}} ||G(\jmath w)|| = \sup_{w \in \mathbb{R}} \sigma_{\max}(G(\jmath w)).$$

Definition 10. *The Hardy space $\mathcal{H}_\infty$ is the function space of matrix-or scalar-valued functions that are analytic and bounded in $\mathbb{C}^+ := \{z \in \mathbb{C} : \mathrm{Re}\,(z) > 0\}$.*

The $\mathcal{H}_\infty$-norm is defined as

$$||F||_\infty := \sup_{\mathrm{Re}\,(s) > 0} \sigma_{\max}(F(s)) = \sup_{w \in \mathbb{R}} \sigma_{\max}(F(\jmath w)).$$

The second equality follows from the maximum modulus theorem.

Definition 11. *The Hardy space $\mathcal{H}_2(\mathbb{C}^+)$ is the function space of matrix-or scalar-valued functions that are analytic in $\mathbb{C}^+$ and bounded w.r.t. the $\mathcal{H}_2$-norm defined as*

$$\begin{aligned}||F||_2 &:= \frac{1}{2\pi}\left(\sup_{\mathrm{Re}\,(\sigma) > 0} \int_{-\infty}^{\infty} ||F(\sigma + \jmath w)||_F^2 dw\right)^{\frac{1}{2}} \\ &= \frac{1}{2\pi}\left(\int_{-\infty}^{\infty} ||F(\jmath w)||_F^2\, dw\right)^{\frac{1}{2}}.\end{aligned} \qquad (2.4)$$

The last equality in (2.4) follows from Theorem 6.

Following [2], for inputs $u(t)$ with $\int_0^\infty ||u(t)||_2^2 dt \leq 1$, the $\mathcal{H}_2$ approximation error gives the following bound

$$\sup_{t > 0} ||y(t) - \hat{y}(t)||_\infty \leq ||G - \hat{G}||_{\mathcal{H}_2}, \qquad (2.5)$$

where G and $\hat{G}$ are original and reduced transfer functions.

For the practical computation of the $\mathcal{H}_2$-norm, the following theorem is used.

Theorem 7.

$$||G||_2^2 = \mathrm{tr}(B^T Q B) = \mathrm{tr}(CPC^T),$$

where P, Q are the controllability and observability Gramians of the corresponding LTI system.

Theorem 8 (Plancherel Theorem). *The Fourier transform of* $f \in L_2^n(-\infty, \infty)$:

$$F(\xi) = \int_{-\infty}^{\infty} f(t)e^{-\xi t}dt$$

is a Hilbert space isomorphism between $L_2^n(-\infty, \infty)$ *and* $\mathcal{L}_2(j\mathbb{R})$. *Furthermore, the Fourier transform maps* $L_2^n(0, \infty)$ *onto* $\mathcal{H}_2(\mathbb{C}^+)$. *In addition it is an isometry, that is, it preserves distances:*

$$L_2^n(-\infty, \infty) \cong \mathcal{L}_2(j\mathbb{R}), \quad L_2^n(0, \infty) \cong \mathcal{H}_2(\mathbb{C}^+).$$

Consequently, the L_2^n-norm in time domain and $\mathcal{L}_2$-norm in frequency domain coincide as well as the $L_2^n(0, \infty)$- and $\mathcal{H}_2$-norms.

Therefore the output error bound (obtained from (2.3)),

$$||y - \hat{y}||_2 = ||Gu - \hat{G}u||_2 \leq ||G - \hat{G}||_\infty ||u||_2, \tag{2.6}$$

holds in time and frequency domain due to the Plancherel theorem, i.e., the $||\cdot||_2$ in (2.6) can be the L_2^n-norm in time domain, or the $\mathcal{L}_2$-norm in frequency domain. Model order reduction aims to compute a reduced-order model such that either $||G - \hat{G}||_\infty < tol$ (2.6) or $||G - \hat{G}||_2 < tol$ (2.5), where tol is an acceptable error tolreance.

3 Methods based on Padé approximation and rational interpolation

The MOR methods based on Padé approximation [4, 11, 13] and rational interpolation [17, 19] are motivated by approximating certain power series expansions of the transfer function by matching a certain number of leading terms. In this section, we consider a slightly generalized system description:

$$\begin{aligned} E\dot{x}(t) &= Ax(t) + Bu(t), \\ y(t) &= Cu(t) + Du(t), \end{aligned} \tag{3.1}$$

where $E \in \mathbb{R}^{n \times n}$ can be singular, and only $\lambda E - A$ is required to be regular, i.e., $\det(\lambda E - A) \not\equiv 0$. A system with possibly singular E is called descriptor system, and its treatment is in general more complex than for the standard state space system in (1.1). But the methods in this section can be represented for singular E without extra effort, therefore we decided to show the results in this generality, in contrast to other methods where the treatment of singular E is much more technically involved.

Methods based on Padé approximation

To consider the transfer function $G(s)$, for simplicity and without loss of generality, we let $D = 0$. (All results hold for nonzero D if in the reduced-order model, $\hat{D} = D$ is used.) Let $s = s_0 + \sigma$, then within the convergence radius of the series,

$$
\begin{aligned}
G(s_0 + \sigma) &= C[(s_0 + \sigma)E - A]^{-1}B \\
&= C[\sigma E + (s_0 E - A)]^{-1}B \\
&= C[I + \sigma(s_0 E - A)^{-1}E]^{-1}[(s_0 E - A)]^{-1}B \\
&= C[I - \sigma(s_0 E - A)^{-1}E + \sigma^2[(s_0 E - A)^{-1}E]^2 - \ldots](s_0 E - A)^{-1}B \\
&= \sum_{i=0}^{\infty} \underbrace{C[-(s_0 E - A)^{-1}E]^i (s_0 E - A)^{-1}B}_{:=m_i(s_0)}\, \sigma^i,
\end{aligned}
$$

where $m_i(s_0)$, $i = 0, 1, 2, \ldots$ are called the *moments* of the transfer function. Note that the series expansion follows from the Neumann Lemma. If the expansion point is chosen as $s_0 = 0$, then the moments are simply $m_i(0) = -C(A^{-1}E)^i A^{-1}B$. For $s_0 = \infty$ and $E = I$, the moments are also called Markov parameters, $m_i(\infty) = CA^i B$. (The $s_0 = \infty$ case for singular E is treated in [7].) In fact, for $s_0 < \infty$, the moments $m_i(s_0)$ are nothing but the ith derivative of $G(s)$ at s_0, multiplied with an appropriate scalar $\frac{1}{i!}$.

The projection matrices $V \in \mathbb{R}^{n \times r}$ and $W \in \mathbb{R}^{n \times r}$ are computed from the moments $m_i(s_0)$, using

$$
\begin{aligned}
\text{range}\{V\} &\supset \text{span}\{\tilde{B}(s_0),\ \tilde{A}(s_0)\tilde{B}(s_0),\ \ldots,\ \tilde{A}^{q-1}(s_0)\tilde{B}(s_0)\}, \\
\text{range}\{W\} &\supset \text{span}\{\tilde{C}(s_0),\ \tilde{A}^T(s_0)\tilde{C}(s_0),\ \ldots,\ (\tilde{A}^T(s_0))^{q-1}\tilde{C}(s_0)\},
\end{aligned}
\tag{3.2}
$$

where $\tilde{A}(s_0) = (s_0 E - A)^{-1}E$, $\tilde{B}(s_0) = (s_0 E - A)^{-1}B$, $\tilde{C}(s_0) = E^T(s_0 E - A)^{-T}C^T$ and $q \ll n$. The following theorem shows that the transfer function of the reduced model computed by the above V and W interpolates the transfer function of the original system up to the $(2q\text{-}1)$th derivative of $G(s)$ at s_0 [11].

Theorem 9. *If the columns of W and V generate bases of the subspaces in (3.2), then the transfer function $\hat{G}(s)$ of the reduced model matches the first $2q$ moments of the transfer function of the original system, i.e.,*

$$
m_i(s_0) = \hat{m}_i(s_0), \quad i = 0, 1, \ldots, 2q - 1,
$$

where $\hat{m}_i(s_0) = \hat{C}[-(s_0 \hat{E} - \hat{A})^{-1}\hat{E}]^i (s_0 \hat{E} - \hat{A})^{-1}\hat{B}$ for $i = 0, 1, \ldots, 2q-1$ are the ith order moments of $\hat{G}$ and $\hat{E} = W^T V$.

It is shown in [11] that the transfer function $\hat{G}(s)$ is a Padé approximant [3] of $G(s)$ for SISO systems. Note the order r of the reduced system may differ from qm, the number of vectors contained in the right-hand side expressions in (3.2) due to deflation occurring when linear dependencies arise. This becomes more likely to happen when m is getting large. Also, to have the same number of columns in V and W, there might be some adaptations necessary depending on the sizes of m and p, and the number of deflated vectors.

Methods based on rational interpolation

Instead of using a single expansion point, multiple expansion points can be used to have multiple series expansions of $G(s)$ about expansion points s_i, $i = 1, \ldots, k$. The matrices V, W can be computed by the combined Krylov subspaces for each s_i, e.g.

$$
\begin{aligned}
\text{range}(V) &\supset \bigcup_{i=1}^{k} \mathcal{K}_{q_i}(\tilde{A}_B(s_i), \tilde{B}(s_i)), \\
\text{range}(W) &\supset \bigcup_{i=1}^{k} \mathcal{K}_{q_i}(\tilde{A}_C(s_i), \tilde{C}(s_i)),
\end{aligned}
\tag{3.3}
$$

where $\tilde{A}_B(s_i) = (s_i E - A)^{-1} E$, $\tilde{A}_C(s_i) = (s_i E - A)^{-T} E^T$, $\tilde{C}(s_i) = (s_i E - A)^{-T} C^T$, and $\mathcal{K}_{q_i}(M, R)$ is the block Krylov subspace $\mathcal{K}_{q_i}(M, R) = \{R, MR, \ldots, M^{q_i-1} R\}$ generated by a square matrix $M \in \mathbb{R}^{n \times n}$ and a rectangular matrix $R \in \mathbb{R}^{n \times n_R}$.

The resulting reduced model matches the first $2q_i$ moments $m_0(s_i)$, $\ldots, m_{2q_i-1}(s_i)$ at each s_i, $i = 1, \ldots, k$, [17]. In other words, the transfer function $\hat{G}(s)$ interpolates $G(s)$ at s_j, $j = 1, \ldots, k$, up to the $(2q_i - 1)$th order derivative. Notice that the starting matrix (vector) for W in (3.3) is $\tilde{C}(s_i) = (s_i E - A)^{-T} C^T$, and $\tilde{A}_C(s_i)$ is not the transpose of $\tilde{A}_B(s_i)$.

For SISO systems, B and C^T are vectors and the matrices V, W in (3.2) can be simultaneously computed by the Lanczos algorithm [11], such that $W^T V = I$, i.e., the columns of W are biorthogonal to the columns of V. For a system with multiple inputs and multiple outputs, B and C are matrices, then the block or band Lanczos algorithms are used in [12, 13, 14] to compute V and W in (3.2).

If only the matrix V is used to compute the reduced model, i.e., $W = V$, then the Arnoldi process can be applied to compute V in (3.2) for a SISO system, and the band Arnoldi process [13] can be applied to compute V for a MIMO system. In this case, in general the number of matched moments/interpolated derivatives is halved. For more discussions on the algorithms for computing V, W in (3.2), see [4, 13]. In [17], algorithms for computing V and W in (3.3) are discussed in detail.

Nowadays, attention is paid to automatic generation of the reduced model. For the methods based on rational interpolation, the question is how to adaptively select the interpolation points s_j, $j = 0, \ldots, k$. Many techniques have been proposed so far, though most of them are more or less heuristic. The algorithm IRKA proposed in [19] iteratively selects the interpolation points s_j, so that upon convergence, the necessary conditions for a locally optimal reduced model w.r.t. the H_2-norm are satisfied for SISO systems. For MIMO systems, tangential interpolation conditions can be used so that IRKA is also able to compute locally H_2-optimal reduced order models in this case [19].

4   Modal truncation

In this section, we again consider only the standard state space system in (1.1).

The modal truncation method [10] is based on the eigendecomposition of the system matrix A in (1.1). Assume that A is diagonalizable, i.e. $T^{-1}AT = D_A$ (T can be a complex matrix), then the matrices V, W for the reduced model are constructed as

$$V = T(:, 1:r) = [t_1, \ldots, t_r],$$
$$\tilde{W}^* = T^{-1}(1:r, :), \quad W = \tilde{W}(V^*\tilde{W})^{-1}.$$

Here, the columns in $T = [t_1, \ldots, t_n]$ are eigenvectors of A, $D_A = \text{diag}(\lambda_1, \ldots, \lambda_n)$ contains the eigenvalues of A. The matrix V is composed of the first r dominant eigenvectors of A, corresponding to the eigenvalues closest to the imaginary axis. The necessary partial eigendecomposition of A can be computed by, e.g. Krylov subspace methods or the Jacobi-Davidson method. That is, only the dominant eigenvectors are computed rather than the full spectral decomposition.

The reduced model is given by $\hat{A} = W^*AV = \text{diag}(\lambda_1, \ldots, \lambda_r)$, $\hat{B} = W^*B$, $\hat{C} = CV$. This is equivalent to doing truncation for the following matrices,

$$T^{-1}AT = \begin{bmatrix} \hat{A} & \\ & \hat{A}_2 \end{bmatrix}, \quad T^{-1}B = \begin{bmatrix} \hat{B} \\ \hat{B}_2 \end{bmatrix}, \quad CT = [\hat{C}, \hat{C}_2].$$

The error bound for the transfer function of the reduced model is

$$\|G - \hat{G}\|_\infty \le \|C_2\| \|B_2\| \frac{1}{\min_{\lambda \in \Lambda(\hat{A}_2)} |\operatorname{Re}(\lambda)|}.$$

The error bound is not computable for very large-scale systems, since the whole spectrum of A needs to be computed in principle.

The modal truncation method only takes into account information from A, the information from B and C is not considered, which can result in a poor approximation of the system's dynamics. An improvement over this is achieved by the dominant pole algorithm [29], where A, B and C are used to measure the dominant poles. The left and right eigenvectors corresponding to the dominant poles are used to construct the reduced model.

5 Balanced truncation

The balanced truncation method was proposed in [26]. The basic principle of balanced truncation is as follows. Firstly, the Gramian matrices P and Q are computed by solving the Lyapunov equations in (2.2). Secondly, a balancing matrix $T = T_b$ (see Theorem 3) is used to obtain a balanced system by state space transformation, $\tilde{A} = TAT^{-1}$, $\tilde{B} = TB$, $\tilde{C} = CT^{-1}$. It can be readily verified that the Gramians of the transformed system are diagonal matrices, i.e. $TPT^T = \Sigma$, $T^{-T}QT^{-1} = \Sigma$.

If $\sigma_{r+1} \ll \sigma_r$, $\Sigma = \text{diag}(\sigma_1, \ldots, \sigma_n)$, can be divided into two parts $\Sigma_1 = \text{diag}(\sigma_1, \ldots, \sigma_r)$, $\Sigma_2 = \text{diag}(\sigma_{r+1}, \ldots, \sigma_n)$. According to this separation, $\tilde{A}$, $\tilde{B}$ and $\tilde{C}$ can be divided as

$$\tilde{A} = \begin{bmatrix} A_{11} & A_{12} \\ A_{21} & A_{22} \end{bmatrix}, \quad \tilde{B} = \begin{bmatrix} B_1 \\ B_2 \end{bmatrix}, \quad \tilde{C} = [C_1, C_2],$$

where $A_{11} \in \mathbb{R}^{r \times r}$, $B_1 \in \mathbb{R}^{r \times m}$, $C_1 \in \mathbb{R}^{p \times r}$ correspond to Σ_1. The reduced model is constructed as $\hat{A} = A_{11}$, $\hat{B} = B_1$, $\hat{C} = C_1$, $\hat{D} = D$.

The motivation of balanced truncation is that the HSVs are the invariants of the system, which means HSVs do not change under state space transformation. Once a system is balanced, the smallest HSVs can be easily read off the diagonalized Gramian Σ, and the system can be truncated according to the separation of Σ. With the deletion of the smallest HSVs, the unimportant states which are difficult to observe and difficult to control are truncated from the system [1], so that only important information of the original system is retained in the reduced model.

In practice, the reduced model is obtained not by explicitly forming the balanced system, instead, the square root (SR) method is used to compute the balanced reduced model. The basic idea is to use the SVD decomposition of SR^T,

$$SR^T = [U_1, U_2] \begin{bmatrix} \Sigma_1 & \\ & \Sigma_2 \end{bmatrix} \begin{bmatrix} V_1^T \\ V_2^T \end{bmatrix}.$$

The two matrices V and W are computed as $W = R^T V_1 \Sigma_1^{-\frac{1}{2}}$, $V = S^T U_1 \Sigma_1^{-\frac{1}{2}}$. The reduced system matrices are $\hat{A} = W^T A V$, $\hat{B} = W^T B$,

$\hat{C} = CV$. It is easily verified that $W^T V = I$, so that VW^T is an oblique projector, hence balanced truncation is a Petrov-Galerkin projection method.

An important property of balanced truncation is the computable error bound,

$$\|G - \hat{G}\|_\infty \leq 2 \sum_{j=r+1}^{n} \sigma_j. \tag{5.1}$$

Thus, from (2.6), we get

$$\|y - \hat{y}\|_2 \leq \left(2 \sum_{j=r+1}^{n} \sigma_j \right) \|u\|_2.$$

From the error bound, the reduced model can be automatically obtained by adaptively choosing r according to the desired accuracy. The properties of the reduced model computed by balanced truncation are summarized in the following theorem.

Theorem 10. *Let the reduced-order system $\hat{\Sigma} : (\hat{A}, \hat{B}, \hat{C}, \hat{D})$ with $r \leq n$ be computed by balanced truncation. Then the reduced-order model $\hat{\Sigma}$ is balanced, stable, minimal, and its HSVs are $\sigma_1, \ldots, \sigma_r$.*

6   Balancing related methods

The balancing related methods were developed for different purposes of model reduction. The linear-quadratic Gaussian balanced truncation (LQGBT) method in [21] can be used as a model reduction method for unstable systems, and it also provides a closed-loop balancing technique. Compared with the standard balanced truncation method in Section 5, the only difference is that the controllability and the observability Gramians are replaced by the solutions P, Q of the dual algebraic Riccati equations (AREs)

$$AP + PA^T - PC^T CP + BB^T = 0,$$
$$A^T Q + QA - QBB^T Q + C^T C = 0.$$

The stochastic balancing method (BST) firstly appeared in [9] for balancing stochastic systems, and was generalized in [15], where a relative error bound for the reduced model is proposed. Instead of solving two Lyapunov equations required by the standard balanced truncation method, one Laypunov equation and one ARE must be solved to get the Gramians P and Q,

$$AP + PA^T + BB^T = 0,$$
$$\bar{A}^T Q + Q\bar{A} + QB_W (DD^T)^{-1} B_W^T Q + C^T (DD^T)^{-1} C = 0,$$

where $\bar{A} := A - B_W(DD^T)^{-1}C, B_W := BD^T + PC^T$.

The positive real balanced truncation method [9, 16] is applicable for positive real systems, also called passive systems. The method is based on the positive-real equations, related to the positive real (Kalman-Yakubovich-Popov-Anderson) lemma. The following two AREs need to be solved,

$$\bar{A}P + P\bar{A}^T + PC^T\bar{R}^{-1}CP + B\bar{R}^{-1}B^T = 0,$$
$$\bar{A}^TQ + Q\bar{A} + QB\bar{R}^{-1}B^TQ + C^T\bar{R}^{-1}C = 0,$$

where $\bar{A} := A - B\bar{R}^{-1}C, \bar{R} := D + D^T$.

In contrast to the error bound for the standard balanced truncation method in (5.1), the computable error bounds for the LQGBT method and the BST method are

$$\text{LQGBT} : ||[N, M] - [\hat{N}, \hat{M}]||_\infty \le 2 \sum_{j=r+1}^{n} \frac{\sigma_j^{LQG}}{\sqrt{1 + (\sigma_j^{LQG})^2}},$$

where N, M and $\hat{N}, \hat{M}$ are stable co-prime factors of G and $\hat{G}$, respectively. That is, $G(s) = M^{-1}N, \hat{G}(s) = \hat{M}^{-1}\hat{N}$.

$$\text{BST} : ||G - \hat{G}||_\infty \le \left(\prod_{j=r+1}^{n} \frac{1 + \sigma_j^{BST}}{1 - (\sigma_j^{BST})} - 1 \right) ||G||_\infty.$$

Actually, the error bound for the BST method is an error bound for the relative error.

Other balancing-based methods include bounded-real balanced truncation [27], H_∞ balanced truncation [25], as well as frequency-weighted versions of the above approaches. A good textbook for learning the balanced truncation methods is [1], where the mathematical basics required for model reduction are also provided. For a deeper study of modal truncation and details of the dominant pole method, please refer to the thesis [29]. In the thesis [17], methods based on Padé approximation are reviewed, and methods based on rational interpolation are proposed.

7   Solving matrix equations

The major computational part of the balanced truncation methods or the balancing related methods is solving the large-scale matrix equations. The efficiency of these model order reduction methods depends on fast numerical algorithms for solving the matrix equations.

Solvability and complexity issues

Consider the Sylvester equation $AX + XB + W = 0$, $A \in \mathbb{R}^{n \times n}$, $B \in \mathbb{R}^{m \times m}$, $X \in \mathbb{R}^{n \times m}$, $W \in \mathbb{R}^{n \times m}$, using the Kronecker (tensor) product, $AX + XB + W = 0$ is equivalent to

$$((I_m \otimes A) + (B^T \otimes I_n)) \operatorname{vec}(X) = \operatorname{vec}(-W). \tag{7.1}$$

Observing that

$$
\begin{aligned}
M := (I_m \otimes A) + (B^T \otimes I_n) \text{ is invertible} \\
\iff \quad 0 \notin \Lambda(M) = \Lambda((I_m \otimes A) + (B^T \otimes I_n)) \\
= \{\lambda_j + \mu_k, \mid \lambda_j \in \Lambda(A), \ \mu_k \in \Lambda(B)\} \\
\iff \quad \Lambda(A) \cap \Lambda(-B) = \emptyset,
\end{aligned}
$$

we have the following corollary.

Corollary 1. *If A, B are Hurwitz matrices, then the Sylvester equation $AX + XB + W = 0$ has unique solution.*

Note that when $B = A^T$, we get the Lyapunov equation

$$AX + XA^T + W = 0. \tag{7.2}$$

A straightforward way of solving the Sylvester equation is via the equivalent linear system of equations in (7.1). This requires the LU factorization of an $nm \times nm$ matrix; for $n \approx m$, the computational complexity is $\frac{2}{3}n^6$. The storage memory is also unacceptable, since we need n^4 data for X.

Traditional methods of solving the matrix equations include the Bartels-Stewart method for Sylvester and Lyapunov equations, the Hessenberg-Schur method for Sylvester equations, and Hammarling's method for Lyapunov equations with A Hurwitz.

All methods are based on the fact that if A, B^T are in Schur form, then

$$M = (I_m \otimes A) + (B^T \otimes I_n)$$

is block-upper triangular. Hence, $Mx = b$ can be solved by back-substitution.

However, clever implementation of the back-substitution process still requires $nm(n + m)$ flops. All methods require the Schur decomposition of A and/or Schur or Hessenberg decomposition of B, which requires $25n^3$ flops for Schur decomposition. Therefore, these methods are not feasible for large-scale problems with $n > 10,000$.

The sign function method

The sign function method is used to solve the Lyapunov equation in (7.2).

Definition 12. *For $Z \in \mathbb{R}^{n \times n}$ with $\Lambda(Z) \cap \imath \mathbb{R} = \emptyset$ and Jordan canonical form*

$$Z = S \begin{bmatrix} J^+ & 0 \\ 0 & J^- \end{bmatrix} S^{-1},$$

the matrix sign function is

$$\operatorname{sign}(Z) := S \begin{bmatrix} I_k & 0 \\ 0 & -I_{n-k} \end{bmatrix} S^{-1}.$$

Lemma 5. *Let $T \in \mathbb{R}^{n \times n}$ be nonsingular and Z as above, then*

$$\operatorname{sign}(TZT^{-1}) = T\operatorname{sign}(Z)T^{-1}.$$

Since $\operatorname{sign}(Z)$ is the square root of I_n, i.e., $(\operatorname{sign}(Z))^2 - I_n = 0$, one can use Newton's method to get $\operatorname{sign}(Z)$ by solving $f(\tilde{Z}) := \tilde{Z}^2 - I_n = 0$:

$$\tilde{Z}_0 \leftarrow Z, \quad \tilde{Z}_{j+1} \leftarrow \frac{1}{2}\left(c_j \tilde{Z}_j + \frac{1}{c_j}\tilde{Z}_j^{-1}\right), \quad j = 1, 2, \ldots, \qquad (7.3)$$

finally, $\operatorname{sign}(Z) = \lim_{j \to \infty} \tilde{Z}_j$ [20]. The variable $c_j > 0$ is a scaling parameter for convergence acceleration and rounding error minimization, e.g.

$$c_j = \sqrt{\frac{\|\tilde{Z}_j^{-1}\|_F}{\|\tilde{Z}_j\|_F}},$$

based on "equilibrating" the norms of the two summands.

Solving the Lyapunov equation in (7.2) with the matrix sign function method is based on the following observation. If $X \in \mathbb{R}^{n \times n}$ is a solution of (7.2), then

$$\underbrace{\begin{bmatrix} I_n & -X \\ 0 & I_n \end{bmatrix}}_{=T^{-1}} \underbrace{\begin{bmatrix} A & W \\ 0 & -A^T \end{bmatrix}}_{=:H} \underbrace{\begin{bmatrix} I_n & X \\ 0 & I_n \end{bmatrix}}_{=:T} = \begin{bmatrix} A & 0 \\ 0 & -A^T \end{bmatrix}.$$

Hence, if A is Hurwitz (i.e., asymptotically stable), then

$$\operatorname{sign}(H) = \operatorname{sign}\left(T \begin{bmatrix} A & 0 \\ 0 & -A^T \end{bmatrix} T^{-1}\right) = T\operatorname{sign}\left(\begin{bmatrix} A & 0 \\ 0 & -A^T \end{bmatrix}\right)T^{-1}$$
$$= \begin{bmatrix} -I_n & 2X \\ 0 & I_n \end{bmatrix}.$$

Applying the sign function iteration in (7.3): $\tilde{Z} \leftarrow \frac{1}{2}(\tilde{Z} + \tilde{Z}^{-1})$ $(c_j = 1)$ using $\tilde{Z}_0 = H = \begin{bmatrix} A & W \\ 0 & -A^T \end{bmatrix}$, and observe that

$$H + H^{-1} = \begin{bmatrix} A & W \\ 0 & -A^T \end{bmatrix} + \begin{bmatrix} A^{-1} & A^{-1}WA^{-T} \\ 0 & -A^{-T} \end{bmatrix},$$

we get the sign function iteration for the Lyapunov equation:

$$\begin{array}{ll} A_0 \leftarrow A, & A_{j+1} \leftarrow \frac{1}{2}\left(A_j + A_j^{-1}\right), \\ W_0 \leftarrow W, & W_{j+1} \leftarrow \frac{1}{2}\left(W_j + A_j^{-1}W_j A_j^{-T}\right), \end{array} \qquad j = 0, 1, 2, \ldots. \quad (7.4)$$

Define $A_\infty := \lim_{j \to \infty} A_j$, $W_\infty := \lim_{j \to \infty} W_j$, we immediately get the following theorem.

Theorem 11. *If A is Hurwitz, then*

$$A_\infty = -I_n \quad and \quad X = \frac{1}{2}W_\infty.$$

Now consider the second iteration in (7.4) for $W_j = B_j B_j^T$, starting with $W_0 = BB^T =: B_0 B_0^T$, one can see that

$$\begin{aligned} \frac{1}{2}\left(W_j + A_j^{-1}W_j A_j^{-T}\right) &= \frac{1}{2}\left(B_j B_j^T + A_j^{-1}B_j B_j^T A_j^{-T}\right) \\ &= \frac{1}{2}\left[B_j \ A_j^{-1}B_j\right]\left[B_j \ A_j^{-1}B_j\right]^T. \end{aligned}$$

Hence, the factored iteration for the sign function method is [8],

$$B_{j+1} \leftarrow \frac{1}{\sqrt{2}}\left[B_j \ A_j^{-1}B_j\right] \qquad (7.5)$$

with $S := \frac{1}{\sqrt{2}}\lim_{j \to \infty} B_j$ and $X = SS^T$. From Theorem 11, a simple stopping criterion is taken as $\|A_j + I_n\|_F \leq tol$. It is clear that the iteration in (7.5) can be used to solve the Lyapunov equations in (2.2) to get the controllability and observability Gramians P and Q.

The alternating direction implicit (ADI) method

The Peaceman Rachford ADI method was originally used to solve the linear system $Au = b$, where $A \in \mathbb{R}^{n \times n}$ is symmetric positive definite and $b \in \mathbb{R}^n$. The idea is to decompose $A = H + V$ with $H, V \in \mathbb{R}^{n \times n}$ such that

$$\begin{aligned} (H + pI)v &= r, \\ (V + pI)w &= t \end{aligned}$$

can be solved easily or efficiently. The standard ADI iteration for solving $Au = b$ is as follows. If H, V are symmetric positive definite matrices, then $\exists p_k$, $k = 1, 2, \ldots$ such that

$$
\begin{aligned}
u_0 &= 0, \\
(H + p_k I)u_{k-\frac{1}{2}} &= (p_k I - V)u_{k-1} + b, \\
(V + p_k I)u_k &= (p_k I - H)u_{k-\frac{1}{2}} + b
\end{aligned}
$$

converges to $u \in \mathbb{R}^n$ solving $Au = b$.

Notice that the Lyapunov operator $\mathcal{L} : P \mapsto AX + XA^T$ can be decomposed into the linear operators,

$$
\mathcal{L}_H : X \mapsto AX, \quad \mathcal{L}_V : X \mapsto XA^T.
$$

In analogy to the standard ADI method, we find the ADI iteration for the Lyapunov equation $AX + XA^T + W = 0$ [32],

$$
\begin{aligned}
X_0 &= 0, \\
(A + p_k I)X_{k-\frac{1}{2}} &= -W - X_{k-1}(A^T - p_k I), \\
(A + p_k I)X_k^T &= -W - X_{k-\frac{1}{2}}^T(A^T - p_k I).
\end{aligned}
$$

Consider applying the above ADI iteration to the Lyapunov equation $AX + XA^T + BB^T = 0$ for a stable matrix $A \in \mathbb{R}^{n \times n}$, with $B \in \mathbb{R}^{n \times m}$, $m \ll n$. The two-step ADI iteration can be rewritten into one step by removing $X_{k-\frac{1}{2}}$,

$$
\begin{aligned}
Z_0 Z_0^T &= 0, \\
Z_k Z_k^T &= -2p_k(A + p_k I)^{-1} BB^T (A + p_k I)^{-T} \\
&\quad + (A + p_k I)^{-1}(A - p_k I)Z_{k-1}Z_{k-1}^T(A - p_k I)^T(A + p_k I)^{-T},
\end{aligned}
$$

with the low-rank factorization of X_k, $X_k = Z_k Z_k^T$, $k = 0, \ldots, k_{\max}$, $Z_k \in \mathbb{R}^{n \times r_k}$, $r_k \ll n$. This is the scheme of the low-rank (vector) ADI method [5, 22, 18, 28]. From the above iteration for $Z_k Z_k^T$, it is easily known that the low-rank factor Z_k of X_k can be iteratively computed as

$$
Z_k = [\sqrt{-2p_k}(A + p_k I)^{-1}B, \ (A + p_k I)^{-1}(A - p_k I)Z_{k-1}],
$$

so that in practical implementations only Z_k is iterated.

Notice that at each iteration step k, the number of vectors needing to be updated in Z_k increases by m. A more efficient algorithm for computing Z_k is proposed in [22], which keeps the number of updated vectors constant at each iteration step.

Assuming $k_{\max}$ is the maximal number of iterations, and observing that $(A - p_i I)$, $(A + p_k I)^{-1}$ commute, then at the last step $k_{\max}$, $Z_{k_{\max}-1}$ can be rewritten as (see [22]),

$$
\begin{aligned}
Z_{k_{\max}-1} = [z_{k_{\max}}, \ P_{k_{\max}-1}z_{k_{\max}}, \ P_{k_{\max}-2}(P_{k_{\max}-1}z_{k_{\max}}), \ \ldots, \\
\ldots, P_1(P_2 \cdots P_{k_{\max}-1}z_{k_{\max}})] ,
\end{aligned} \tag{7.6}
$$

$$z_{k_{max}} = \sqrt{-2p_{k_{max}}}(A + p_{k_{max}}I)^{-1}B$$

and

$$P_i := \frac{\sqrt{-2p_i}}{\sqrt{-2p_{i+1}}} \left[I - (p_i + p_{i+1})(A + p_iI)^{-1}\right].$$

From (7.6), we derive the iteration for Z_k, $k = 0, 1, \ldots, k_{max} - 1$,

$$\begin{aligned} Z_0 &= z_{k_{max}}, \\ Z_k &= [z_{k_{max}}, P_{k_{max}-1}z_{k_{max}}, \ldots, P_{k_{max}-k}\cdots(P_{k_{max}-1}z_{k_{max}})], \end{aligned} \tag{7.7}$$

where the number of updated vectors at each step is always m.

Factored Galerkin method

The factored Galerkin method is a projection-based method for solving Lyapunov equations with $A + A^T < 0$. The basic steps are

1. Compute an orthonormal basis $Z = [z_1, \ldots, z_r] \in \mathbb{R}^{n \times r}$ of the subspace $\mathcal{Z} \subset \mathbb{R}^n$ ($\dim \mathcal{Z} = r$), i.e., $\text{range}(Z) = \mathcal{Z}$.
2. Set $A_r := Z^T A Z$, $B_r := Z^T B$.
3. Solve small-size Lyapunov equation $A_r \hat{X} + \hat{X} A_r^T + B_r B_r^T = 0$.
4. Use $X \approx Z \hat{X} Z^T$.

The subspace $\mathcal{Z}$ can be taken as, e.g.

$$\mathcal{Z} = \mathcal{K}_r(A, B) = \text{span}\{B, AB, A^2B, \ldots, A^{r-1}B\},$$

which corresponds to the Krylov subspace methods.

The K-PIK method [30] uses the combined subspace

$$\mathcal{Z} = \mathcal{K}_r(A, B) \cup \mathcal{K}_r(A^{-1}, A^{-1}B).$$

The rational Krylov subspace method uses

$$\mathcal{Z} = \text{colspan}\{(A - s_1I)^{-1}B, \ldots, (A - s_rI)^{-1}B\}.$$

$\mathcal{Z}$ can also be taken as the ADI subspace

$$\mathcal{Z} = \text{colspan}\{z_{k_{max}}, P_{k_{max}-1}z_{k_{max}}, \ldots, P_{k_{max}-r+1}\cdots(P_{k_{max}-1}z_{k_{max}})\}.$$

The ADI subspace is proved to be a rational Krylov subspace in [22]. In the following subsections, we discuss the numerical methods for solving large-scale algebraic Riccati equation (ARE)

$$A^T X + X A - X B B^T X + C^T C = 0.$$

Newton's method for AREs

Consider the ARE,

$$0 = \mathcal{R}(X) := A^T X + XA - XBB^T X + C^T C, \qquad (7.8)$$

with $A \in \mathbb{R}^{n \times n}$, $B \in \mathbb{R}^{n \times m}$, $C \in \mathbb{R}^{p \times n}$. The Frechét derivative of $\mathcal{R}(X)$ at X is $\mathcal{R}'_X : Z \mapsto (A - BB^T X)^T Z + Z(A - BB^T X)$.

The Newton-Kantorovich method follows the iteration $X_{j+1} = X_j - (\mathcal{R}'_{X_j})^{-1} \mathcal{R}(X_j)$, $j = 0, 1, 2, \ldots$, and can be described by Algorithm 1.

Algorithm 1 Newton's method for AREs

1: FOR $j = 0, 1, \ldots$
2: $A_j \leftarrow A - BB^T X_j =: A - BK_j$.
3: Solve the Lyapunov equation $A_j^T N_j + N_j A_j = -\mathcal{R}(X_j)$.
4: $X_{j+1} \leftarrow X_j + N_j$.
5: ENDFOR j

If $A_j = A - BK_j = A - BB^T X_j$ is stable $\forall\, j \geq 0$, then $\mathcal{R}(X_j)$ converges to zero, $\lim_{j \to \infty} \|\mathcal{R}(X_j)\|_F = 0$, so X_j converges to the solution of the ARE, $\lim_{j \to \infty} X_j = X_* \geq 0$. It is seen that during the algorithm, large-scale Lyapunov equations need to be efficiently solved, where the algorithms discussed in the above two subsections can be applied.

Low-Rank Newton-ADI for AREs

If we rewrite Newton's method for AREs, in particular Step 3 in Algorithm 1, we get

$$A_j^T \underbrace{(X_j + N_j)}_{=X_{j+1}} + \underbrace{(X_j + N_j)}_{=X_{j+1}} A_j = \underbrace{-C^T C - X_j BB^T X_j}_{=:-W_j W_j^T}.$$

Setting $X_j = Z_j Z_j^T$ for rank$(Z_j) \ll n$, we have

$$A_j^T \left(Z_{j+1} Z_{j+1}^T \right) + \left(Z_{j+1} Z_{j+1}^T \right) A_j + W_j W_j^T = 0. \qquad (7.9)$$

Then $Z_{j+1}, j = 0, 1, \ldots$, can be obtained by solving the Lyapunov equations in (7.9) with the factored ADI iteration in (7.7), so that Algorithm 1 is combined with the low-rank ADI methods.

8   Software

In the toolbox LYAPACK, there are MATLAB routines for solving large, sparse Lyapunov equations, and AREs equations. The main methods used are low-rank ADI and Newton-ADI iterations. It can be downloaded from [34].

The Matrix Equations and Sparse Solvers library (MESS) [35] is the extended and revised version of the LYAPACK Toolbox. It includes solvers for large-scale differential Riccati equations. There are many algorithmic improvements, for example, new ADI parameter selection, column compression, a more efficient use of direct solvers, treatment of generalized systems without factorization of the mass matrix, new ADI versions avoiding complex arithmetic etc. It is available as a MATLAB toolbox, as well as a C-library. The C version provides a large set of auxiliary subroutines for sparse matrix computations and efficient usage of modern multicore workstations.

9 Conclusions

In this chapter, popular model order reduction methods applicable for non-parametric LTI systems are discussed. Numerical algorithms for solving large-scale matrix equations are explored.

Most of the above discussed methods can be either directly applied, or extended to treating descriptor systems $E\dot{x} = Ax + Bu$, E singular. Some methods have been generalized for bilinear and stochastic systems. Methods based on Padé approximation/rational interpolation have also been extended for nonlinear systems. The well-known proper orthogonal decomposition (POD) method for nonlinear systems is not covered in the lecture. Parametric model order reduction (PMOR) for parametric systems, such as

$$\dot{x} = A(p)x + B(p)u, \quad y = C(p)x,$$

where $p \in \mathbb{R}^d$ is a free parameter vector, is a huge topic, and cannot be included here either. For a survey of PMOR methods, see e.g. [6]. For a wide view about various MOR methods for different complex systems, please visit the MORwiki [36], where one is also free to share the benchmarks for testing and verifying MOR algorithms.

References

[1] A. C. Antoulas. Approximation of Large-Scale Dynamical Systems. *SIAM Publications*, Philadelphia, PA, 2005.

[2] A. C. Antoulas, C. A. Beattie, and S. Gugercin. Interpolatory model reduction of large-scale dynamical systems, in Efficient Modeling and Control of Large-Scale Systems, Javad Mohammadpour and Karolos M. Grigoriadis, eds., *Springer* US, 3–58, 2010.

[3] G. A. Baker, Jr. and P. Graves-Morris. Padé Approximants, Second Edition. *Cambridge University Press*, New York, 1996.

[4] Z. Bai. Krylov subspace techniques for reduced-order modeling of large-scale dynamical systems. *Applied Numerical Mathematics*, 43(1–2): 9–44, 2002.

[5] P. Benner, J.-R. Li, and T. Penzl. Numerical Solution of Large Lyapunov Equations, Riccati Equations, and Linear-Quadratic Control Problems. *Numerical Algorithms*, 15(9): 755–777, 2008.

[6] P. Benner, S. Gugercin, and K. Willcox. A survey of model reduction methods for parametric systems, *MPI Magdeburg Preprints* MPIMD/13-14, August 2013. Available from http://www.mpi-magdeburg.mpg.de/preprints/.

[7] P. Benner and V. Sokolov. Partial realization of descriptor systems. *Systems & Control Letters*, 55(11): 929–938, 2006.

[8] P. Benner and E. S. Quintana-Orti. Solving stable generalized Lyapunov equations with the matrix sign function. *Numerical Algorithms*, 20: 75–100, 1999.

[9] U. B. Desai and D. Pal. A transformation approach to stochastic model reduction. *IEEE Transactions on Automatic Control*, AC-29(12): 1097–1100, 1984.

[10] E. J. Davison. A method for simplifying linear dynamic systems. *IEEE Transactions on Automatic Control*, AC-11(1): 93–101, 1966.

[11] P. Feldmann and R. W. Freund. Efficient linear circuit analysis by Padé approximation via the Lanczos process. *IEEE Trans. Comput.-Aided Des. Integr. Circuits Syst.*, 14(5): 639–649, 1995.

[12] P. Feldmann and R. W. Freund. Reduced-order modeling of large linear subcircuits via a block Lanczos algorithm. *Proc. 32nd ACM/IEEE Design Automation Conf.*, 474–479, 1995.

[13] R. W. Freund. Model reduction methods based on Krylov subspaces. *Acta Numerica*, 12: 267–319, 2003.

[14] R. W. Freund. Krylov subspace methods for reduced-order modeling in circuit simulation. *J. Comp. Appl. Math.*, 123: 395–421, 2000.

[15] M. Green. A relative error bound for balanced stochastic truncation. *IEEE Transactions on Automatic Control*, AC-33: 961–965, 1988.

[16] M. Green. Balanced stochastic realizations. *Linear Algebra and its Applications*, 98: 211–247, 1988.

[17] E. J. Grimme. Krylov projection methods for model reduction. PhD thesis, Univ. Illinois, Urbana-Champaign, 1997.

[18] S. Gugercin, D. C. Sorensen, and A. C. Antoulas. A modified low-rank Smith method for large-scale Lyapunov equations. *Numerical Algorithms*, 32(1): 27–55, 2003.

[19] S. Gugercin, A. C. Antoulas, and C. A. Beattie. $\mathcal{H}_2$ model reduction for large-scale linear dynamical systems. *SIAM J. Matrix Anal. Appl.*, 30(2): 609–638, 2008.

[20] N. J. Higham. Functions of Matrices: Theory and Computation. *SIAM*, 2008.

[21] E. Jonckheere and L. Silverman. A new set of invariants for linear systems–application to reduced order compensator. *IEEE Trans Automat. Control*, AC-28: 953–964, 1983.

[22] J.-R. Li and J. White. Low Rank Solution of Lyapunov Equations. *SIAM J. Matrix Anal. Appl.*, 24(1): 260–280, 2002.

[23] A. J. Laub, M. T. Heath, C. C. Paige and R. C. Ward. Computation of system balancing transformations and other applications of simultaneous diagonalization algorithms. *IEEE Transactions on Automatic Control*, AC-32(2): 115–122, 1987.

[24] Maximum-modulus principle. *Encyclopedia of Mathematics*. http://www.encyclopediaofmath.org/index.php/Maximum-modulus_principle

[25] D. Mustafa and K. Glover. Controller reduction by $\mathcal{H}_\infty$-balanced truncation. *IEEE Transactions on Automatic Control*, 36: 668–682, 1991.

[26] B. C. Moore. Principal component analysis in linear systems: Controllability, observability, and model reduction. *IEEE Transactions on Automatic Control*, AC-26: 17–32, 1981.

[27] P. C. Opdenacker and E. A. Jonckheere. A contraction mapping preserving balanced reduction scheme and its infinity norm error bounds. *IEEE Transactions on Circuits and Systems*, CAS 35: 184–189, 1988.

[28] T. Penzl. A cyclic low-rank Smith method for large sparse Lyapunov equations. *SIAM J. Sci. Comput.*, 21(4): 1401–1418, 2000.

[29] J. Rommes. Methods for eigenvalue problems with applications in model order reduction. PhD thesis, Utrecht University (Netherlands), 2007.

[30] V. Simoncini. A new iterative method for solving large-scale Lyapunov matrix equations. *SIAM Journal on Scientific Computing*, 29(3): 1268–1288, 2007.

[31] M. Tombs and I. Postlethwaite. Truncated balanced realization of a stable non-minimal state-space system. *International Journal of Control*, 46: 1319–1330, 1987.

[32] E. L. Wachspress. Iterative solution of the Lyapunov matrix equation. *Applied Mathematics Letters* 1(1), 87–90, 1988.

[33] E. L. Wachspress. The ADI Model Problem, Windsor, CA, 1995

[34] LYAPACK, http://www.tu-chemnitz.de/sfb393/lyapack/.

[35] MESS, http://svncsc.mpi-magdeburg.mpg.de/trac/messtrac.

[36] http://www.modelreduction.org.

Rayleigh Quotient Based Optimization Methods for Eigenvalue Problems

Ren-Cang Li*

Abstract

Four classes of eigenvalue problems that admit similar min-max principles and the Cauchy interlacing inequalities as the symmetric eigenvalue problem famously does are investigated. These min-max principles pave ways for efficient numerical solutions for extreme eigenpairs by optimizing the so-called Rayleigh quotient functions. In fact, scientists and engineers have already been doing that for computing the eigenvalues and eigenvectors of Hermitian matrix pencils $A - \lambda B$ with B being positive definite, the first class of our eigenvalue problems. But little attention has gone to the other three classes: positive semidefinite pencils, linear response eigenvalue problems, and hyperbolic eigenvalue problems, in part because most min-max principles for the latter were discovered only very recently and some more are being discovered. It is expected that they will drive the effort to design better optimization based numerical methods for years to come.

1 Introduction

Eigenvalue problems are ubiquitous. Eigenvalues explain many physical phenomena well such as vibrations and frequencies, (in)stabilities of dynamical systems, and energy levels in molecules or atoms. This chapter focuses on classes of eigenvalue problems that admit various min-max principles and the Cauchy interlacing inequalities as the symmetric eigenvalue problem famously does [4, 38, 47]. These results make it possible to efficiently calculate extreme eigenpairs of the eigenvalue problems by optimizing associated Rayleigh quotients.

Consider the generalized eigenvalue problem

$$Ax = \lambda Bx, \tag{1.1}$$

---

*Department of Mathematics, University of Texas at Arlington, P.O. Box 19408, Arlington, TX 76019. E-mail: rcli@uta.edu. Supported in part by NSF grants DMS-1115834 and DMS-1317330, and a research gift grant from Intel Corporation.

where both A and B are Hermitian. The first class of eigenvalue problems are those for which B is also positive definite. Such an eigenvalue problem is equivalent to a symmetric eigenvalue problem $B^{-1/2}AB^{-1/2}y = \lambda x$ and thus, not surprisingly, all min-max principles (Courant-Fischer, Ky Fan trace min/max, Wielandt-Lidskii) and the Cauchy interlacing inequalities have their counterparts in this eigenvalue problem. The associated *Rayleigh quotient* is

$$\rho(x) = \frac{x^{\mathrm{H}}Ax}{x^{\mathrm{H}}Bx}. \qquad (1.2)$$

When B is indefinite and even singular, (1.1) is no longer equivalent to a symmetric eigenvalue problem in general and it may even have complex eigenvalues which clearly admit no min-max representations. But if there is a real scalar λ_0 such that $A - \lambda_0 B$ is positive semidefinite, then the eigenvalue problem (1.1) has only real eigenvalues and they admit similar min-max principles and the Cauchy interlacing inequalities [25, 27, 29]. This is the second class of eigenvalue problems and it shares the same Rayleigh quotient (1.2) as the first class. We call a matrix pencil in this class a *positive semidefinite pencil*. Opposite to the concept of a *positive semidefinite matrix pencil*, naturally, is that of a *negative semidefinite matrix pencil* $A - \lambda B$ by which we mean that A and B are Hermitian and there is a real λ_0 such that $A - \lambda_0 B$ is negative semidefinite. Evidently, if $A - \lambda B$ is a negative semidefinite matrix pencil, then $-(A-\lambda B) = (-A)-\lambda(-B)$ is a positive semidefinite matrix pencil because $(-A) - \lambda_0(-B) = -(A - \lambda_0 B)$. Therefore it suffices to only study either positive or negative semidefinite pencils.

The third class of eigenvalue problems is the so-called *linear response eigenvalue problem* or *random phase approximation eigenvalue problem*

$$\begin{bmatrix} 0 & K \\ M & 0 \end{bmatrix} \begin{bmatrix} y \\ x \end{bmatrix} = \lambda \begin{bmatrix} y \\ x \end{bmatrix},$$

where K and M are Hermitian and positive semidefinite matrices and one of them is definite. Any eigenvalue problem in this class can be turned into one in the second class by permuting the first and second block rows to get

$$\begin{bmatrix} M & 0 \\ 0 & K \end{bmatrix} \begin{bmatrix} y \\ x \end{bmatrix} = \lambda \begin{bmatrix} 0 & I \\ I & 0 \end{bmatrix} \begin{bmatrix} y \\ x \end{bmatrix},$$

where I is the identity matrix of apt size. In this sense, the third class is a subclass of the second class, but with block substructures. The associated Rayleigh quotient is

$$\rho(x,y) = \frac{x^{\mathrm{H}}Kx + y^{\mathrm{H}}My}{2|x^{\mathrm{H}}y|}.$$

The first minimization principle for such eigenvalue problems was essentially published by Thouless [50], but more were obtained only very recently [2, 3].

The fourth class of eigenvalue problems is the *hyperbolic quadratic eigenvalue problem*

$$(\lambda^2 A + \lambda B + C)x = 0$$

arising from dynamical systems with friction, where A, B, and C are Hermitian and A is positive definite and

$$(x^H B x)^2 - 4(x^H A x)(x^H C x) > 0 \quad \text{for any nonzero vector } x.$$

The associated Rayleigh quotients are

$$\rho_\pm(x) = \frac{-x^H B x \pm \left[(x^H B x)^2 - 4(x^H A x)(x^H C x)\right]^{1/2}}{2(x^H A x)}.$$

Courant-Fischer type min-max principles were known to Duffin [10] and the Cauchy type interlacing inequalities to Veselić [52]. Other min-max principles (Wielandt-Lidskii type, Ky Fan trace min/max type) are being discovered [28].

In the rest of this chapter, we will explain the steepest descent/ascent methods and nonlinear conjugate gradient methods for the first class of eigenvalue problems, including the incorporation of preconditioning techniques, extending search spaces, and block implementations, in detail but only state min-max principles — old and new — for the other three classes. The interested reader can consult relevant references for the corresponding steepest descent/ascent methods and nonlinear conjugate gradient methods or design his own based on the min-max principles stated.

Notation. Throughout this chapter, $\mathbb{C}^{n \times m}$ is the set of all $n \times m$ complex matrices, $\mathbb{C}^n = \mathbb{C}^{n \times 1}$, and $\mathbb{C} = \mathbb{C}^1$, and similarly $\mathbb{R}^{n \times m}$, $\mathbb{R}^n$, and $\mathbb{R}$ are for their real counterparts. I_n (or simply I if its dimension is clear from the context) is the $n \times n$ identity matrix, and e_j is its jth column. The superscript ".^{T}" and ".^{H}" take transpose and complex conjugate transpose of a matrix/vector, respectively. For a matrix X, $\mathcal{R}(X)$ and $\mathcal{N}(X)$ are the column space and null space of X, respectively.

We shall also adopt MATLAB-like convention to access the entries of vectors and matrices. Let $i : j$ be the set of integers from i to j inclusive. For a vector u and a matrix X, $u_{(j)}$ is u's jth entry, $X_{(i,j)}$ is X's (i,j)th entry; X's submatrices $X_{(k:\ell,i:j)}$, $X_{(k:\ell,:)}$, and $X_{(:,i:j)}$ consist of intersections of row k to row ℓ and column i to column j, row k to row ℓ, and column i to column j, respectively.

For $A \in \mathbb{C}^{n \times n}$, $A \succ 0$ ($A \succeq 0$) means that A is Hermitian and positive (semi-)definite, and $A \prec 0$ ($A \preceq 0$) means $-A \succ 0$ ($-A \succeq 0$).

2 Hermitian pencil $A - \lambda B$ with definite B

In this section, we consider the generalized eigenvalue problem

$$Ax = \lambda Bx, \tag{2.1}$$

where $A, B \in \mathbb{C}^{n \times n}$ are Hermitian with $B \succ 0$. When the equation (2.1) for a scalar $\lambda \in \mathbb{C}$ and $0 \neq x \in \mathbb{C}^n$ holds, λ is called an *eigenvalue* and x a corresponding *eigenvector*. Theoretically, it is equivalent to the standard Hermitian eigenvalue problem

$$B^{-1/2} A B^{-1/2} y = \lambda y. \tag{2.2}$$

Both have the same eigenvalues with eigenvectors related by $y = B^{1/2}x$, where $B^{-1/2} = (B^{-1})^{1/2}$ is the positive definite square root of B^{-1} (also $B^{-1/2} = (B^{1/2})^{-1}$) [5, 19].

Numerically, if it has to be done (usually for modest n, up to a few thousands), the conversion of (2.1) to a standard Hermitian eigenvalue problem is usually accomplished through B's Cholesky decomposition: $B = R^{\mathrm{H}}R$, where R is upper triangular, rather than B's square root which is much more expensive to compute but often advantageous for theoretical investigations. The converted eigenvalue problem is then

$$R^{-\mathrm{H}} A R^{-1} y = \lambda y \tag{2.3}$$

with eigenvectors related by $y = Rx$, and can be solved as a dense eigenvalue problem by LAPACK [1] for modest n.

But calculating the Cholesky decomposition can be very expensive, too, for large n, not to mention possible fill-ins for unstructured sparse B. In this section, we are concerned with Rayleigh Quotient based optimization methods to calculate a few extreme eigenvalues of (2.1).

By the theoretical equivalence of (2.1) to the standard Hermitian eigenvalue problem (2.2) or (2.3), we know that (2.1) has n real eigenvalues and B-orthonormal eigenvectors.

Throughout the rest of this section, $A - \lambda B$ will be assumed a Hermitian matrix pencil of order n with $B \succ 0$, and its eigenvalues, eigenvectors, and eigen-decomposition are given by (2.4).

eigenvalues:	$\lambda_1 \leq \lambda_2 \leq \cdots \leq \lambda_n$, and $\Lambda = \mathrm{diag}(\lambda_1, \lambda_2, \ldots, \lambda_n)$,
B-orthonormal eigenvectors:	$u_1, u_2, \ldots, u_n$, and $U = [u_1, u_2, \ldots, u_n]$,
eigen-decomposition:	$U^{\mathrm{H}} A U = \Lambda$ and $U^{\mathrm{H}} B U = I_n$.

$$\tag{2.4}$$

In what follows, our focus is on computing the first few smallest eigenvalues and their associated eigenvectors. The case for the largest few eigenvalues can be dealt with in the same way by replacing A by $-A$, i.e., considering $(-A) - \lambda B$ instead.

2.1   Basic theory

Given $x \in \mathbb{C}^n$, the **Rayleigh Quotient** for the generalized eigenvalue problem $Ax = \lambda Bx$ is defined by

$$\rho(x) = \frac{x^{\mathrm{H}} A x}{x^{\mathrm{H}} B x}. \tag{2.5}$$

Similarly for $X \in \mathbb{C}^{n \times k}$ with $\mathrm{rank}(X) = k$, the **Rayleigh Quotient Pencil** is

$$X^{\mathrm{H}} A X - \lambda X^{\mathrm{H}} B X. \tag{2.6}$$

Theorem 2.1 collects important min-max results and the Cauchy interlacing inequalities for the eigenvalue problem (2.1). They can be derived via the corresponding results for (2.2) or (2.3), the theoretical equivalence of (2.1) [4, 38, 47].

Theorem 2.1. *Let* $A - \lambda B$ *be a Hermitian matrix pencil of order* n *with* $B \succ 0$.

1. **(Courant-Fischer min-max principles)** *For* $j = 1, 2, \ldots, n$,

$$\lambda_j = \min_{\dim \mathcal{X}=j} \ \max_{x \in \mathcal{X}} \rho(x), \tag{2.7a}$$

$$\lambda_j = \max_{\mathrm{codim}\, \mathcal{X}=j-1} \min_{x \in \mathcal{X}} \rho(x). \tag{2.7b}$$

In particular,

$$\lambda_1 = \min_x \rho(x), \quad \lambda_n = \max_x \rho(x). \tag{2.8}$$

2. **(Ky Fan trace min/max principles)** *For* $1 \le k \le n$,

$$\sum_{i=1}^{k} \lambda_i = \min_{X^{\mathrm{H}} B X = I_k} \mathrm{trace}(X^{\mathrm{H}} A X), \tag{2.9a}$$

$$\sum_{i=n-k+1}^{n} \lambda_i = \max_{X^{\mathrm{H}} B X = I_k} \mathrm{trace}(X^{\mathrm{H}} A X). \tag{2.9b}$$

Furthermore if $\lambda_k < \lambda_{k+1}$, *then* $\mathcal{R}(X) = \mathcal{R}(U_{(:,1:k)})$ *for any minimizing* $X \in \mathbb{C}^{n \times k}$ *for* (2.9a); *if* $\lambda_{n-k} < \lambda_{n-k+1}$, *then* $\mathcal{R}(X) = \mathcal{R}(U_{(:,n-k+1:n)})$ *for any maximizing* $X \in \mathbb{C}^{n \times k}$ *for* (2.9b).

3. **(Cauchy interlacing inequalities)** *Let $X \in \mathbb{C}^{n \times k}$ with* $\text{rank}(X) = k$, *and denote by* $\mu_1 \leq \mu_2 \leq \cdots \leq \mu_k$ *the eigenvalues of the Rayleigh quotient pencil (2.6). Then*

$$\lambda_j \leq \mu_j \leq \lambda_{n-k+j} \quad \text{for } 1 \leq j \leq k. \qquad (2.10)$$

Furthermore if $\lambda_j = \mu_j$ *for* $1 \leq j \leq k$ *and* $\lambda_k < \lambda_{k+1}$, *then* $\mathcal{R}(X) = \mathcal{R}(U_{(:,1:k)})$; *if* $\mu_j = \lambda_{n-k+j}$ *for* $1 \leq j \leq k$ *and* $\lambda_{n-k} < \lambda_{n-k+1}$, *then* $\mathcal{R}(X) = \mathcal{R}(U_{(:,n-k+1:n)})$.

The computational implications of these results are as follows. The equations in (2.8) or (2.9) naturally lead to applications of optimization approaches to computing the first/last few eigenvalues and/or their associated eigenvectors, while the inequalities in (2.10) suggest that judicious choices of X can push μ_j either down to λ_j or up to λ_{n-k+j} for the purpose of computing them.

In pertinent to deflation, i.e., avoiding computing known or already computed eigenpairs, we have the following results.

Theorem 2.2. *Let integer* $1 \leq k < n$ *and* $\xi \in \mathbb{R}$.

1. *We have*

$$\lambda_{k+1} = \min_{x \perp_B u_i, \, 1 \leq i \leq k} \rho(x), \quad \lambda_{k+1} = \max_{x \perp_B u_i, \, k+2 \leq i \leq n} \rho(x),$$

   where $\perp_B$ *stands for B-orthogonality, i.e.,* $x \perp_B y$ *means* $\langle x, y \rangle_B \equiv x^H By = 0$.

2. *Let* $V = [u_1, u_2, \ldots, u_k]$. *The eigenvalues of matrix pencil* $[A + \xi(BV)(BV)^H] - \lambda B$ *are*

$$\lambda_j + \xi \text{ for } 1 \leq j \leq k, \quad \lambda_j \text{ for } k+1 \leq j \leq n$$

   with the corresponding eigenvectors u_j *for* $1 \leq j \leq n$. *In particular,*

$$U^H[A + \xi(BV)(BV)^H]U = \begin{bmatrix} \Lambda_1 + \xi I_k & 0 \\ 0 & \Lambda_2 \end{bmatrix}, \quad U^H BU = I_n,$$

   where $\Lambda_1 = \Lambda_{(1:k,1:k)}$ *and* $\Lambda_2 = \Lambda_{(k+1:n,k+1:n)}$.

The concept of invariant subspace is very important in the standard eigenvalue problem and, more generally, operator theory. In a loose sense, computing a few eigenvalues of a large scale matrix $H \in \mathbb{C}^{n \times n}$ is equivalent to calculating a relevant invariant subspace $\mathcal{X}$ of H, i.e., a subspace $\mathcal{X} \subseteq \mathbb{C}^n$ such that $H\mathcal{X} \subseteq \mathcal{X}$. This concept naturally extends to the generalized eigenvalue problem for $A - \lambda B$ that we are interested in, i.e., A and B are Hermitian and $B \succ 0$.

Definition 2.1. A $\mathfrak{X} \subseteq \mathbb{C}^n$ is called a *generalized invariant subspace* of $A - \lambda B$ if

$$A\mathfrak{X} \subseteq B\mathfrak{X}.$$

Sometimes, it is simply called an *invariant subspace*.

Some important properties of an invariant subspace are summarized in the following theorem whose proof is left as an exercise.

Theorem 2.3. *Let $\mathfrak{X} \subseteq \mathbb{C}^n$ and $\dim \mathfrak{X} = k$, and let $X \in \mathbb{C}^{n \times k}$ be a basis matrix of $\mathfrak{X}$.*

1. *$\mathfrak{X}$ is an invariant subspace of $A - \lambda B$ if and only if there is $A_1 \in \mathbb{C}^{k \times k}$ such that*

$$AX = BXA_1. \tag{2.11}$$

2. *Suppose $\mathfrak{X}$ is an invariant subspace of $A - \lambda B$ and (2.11) holds. Then the following statements are true.*

    (a) *$A_1 = (X^H B X)^{-1}(X^H A X)$ and thus it has the same eigenvalues as $X^H A X - \lambda X^H B X$. If X has B-orthonormal columns, i.e., $X^H B X = I_k$ (one can always pick a basis matrix like this), then $A_1 = X^H A X$ which is also Hermitian.*

    (b) *For any eigenpair $(\hat{\lambda}, \hat{x})$ of A_1: $A_1 \hat{x} = \hat{\lambda}\hat{x}$, $(\hat{\lambda}, X\hat{x})$ is an eigenpair of $A - \lambda B$.*

    (c) *Let $X_\perp \in \mathbb{C}^{n \times (n-k)}$ such that $Z := [X, X_\perp]$ is nonsingular and $X^H B X_\perp = 0$. We have*

$$Z^H A Z = \begin{bmatrix} X^H A X & \\ & X_\perp^H A X_\perp \end{bmatrix}, \ Z^H B Z = \begin{bmatrix} X^H B X & \\ & X_\perp^H B X_\perp \end{bmatrix}.$$

2.2 Rayleigh-Ritz procedure

Theorem 2.3 says that partial spectral information can be extracted from an invariant subspace if known. But an exact invariant subspace is hard to come by in practice. Through computations we often end up with subspaces $\mathfrak{X}$ that

1. are accurate approximate invariant subspaces themselves, or

2. have nearby lower dimensional invariance subspaces.

For the former, it means that $\|AX - BXA_1\|$ is tiny for some matrix A_1, where X is a basis matrix of $\mathfrak{X}$ and $\|\cdot\|$ is some matrix norm. For the latter, it means there is an invariant subspace $\mathfrak{U}$ of a lower dimension than $\mathfrak{X}$ such that the canonical angles from $\mathfrak{U}$ to $\mathfrak{X}$ are all tiny.

The Rayleigh-Ritz procedure is a way to extract approximate spectral information on $A - \lambda B$ for a given subspace that satisfies either one of the two requirements.

Algorithm 2.1 Rayleigh-Ritz procedure

Given a computed subspace $\mathcal{X}$ of dimension ℓ in the form of a basis matrix $X \in \mathbb{C}^{n \times \ell}$, this algorithm computes approximate eigenpairs of $A - \lambda B$.

1: compute the projection matrix pencil $X^H A X - \lambda X^H B X$ which is $\ell \times \ell$;

2: solve the eigenvalue problem for $X^H A X - \lambda X^H B X$ to obtain its eigenpairs $(\hat{\lambda}_i, \hat{x}_i)$ which yield approximate eigenpairs $(\hat{\lambda}_i, X\hat{x}_i)$, called *Rayleigh-Ritz pairs*, for the original pencil $A - \lambda B$. These $\hat{\lambda}_i$ are called *Ritz values* and $X\hat{x}_i$ *Ritz vectors*.

If $\mathcal{X}$ is a true invariant subspace, the *Ritz values* and *Ritz vectors* as rendered by this Rayleigh-Ritz procedure are exact eigenvalues and eigenvectors of $A - \lambda B$ in the absence of roundoff errors, as guaranteed by Theorem 2.3. So in this sense, this Rayleigh-Ritz procedure is a natural thing to do. On the other hand, as for the standard symmetric eigenvalue problem, the procedure retains several optimality properties as we shall now explain.

By Theorem 2.1,

$$\lambda_i = \min_{\substack{\mathcal{Y} \subseteq \mathbb{C}^n \\ \dim \mathcal{Y} = i}} \max_{y \in \mathcal{Y}} \rho(y), \tag{2.12}$$

where the minimization is taken over all $\mathcal{Y} \subset \mathbb{C}^n$ with $\dim \mathcal{Y} = i$. So given $\mathcal{X} \subset \mathbb{C}^n$, the natural definition of the best approximation α_i to λ_i is to replace $\mathbb{C}^n$ by $\mathcal{X}$ to get

$$\alpha_i = \min_{\substack{\mathcal{Y} \subseteq \mathcal{X} \\ \dim \mathcal{Y} = i}} \max_{y \in \mathcal{Y}} \rho(y). \tag{2.13}$$

Any $\mathcal{Y} \subseteq \mathcal{X}$ with $\dim \mathcal{Y} = i$ can be represented by its basis matrix $Y \in \mathbb{C}^{n \times i}$ which in turn can be uniquely represented by $\hat{Y} \in \mathbb{C}^{\ell \times i}$ with $\operatorname{rank}(\hat{Y}) = i$ such that $Y = X\hat{Y}$. So $y \in \mathcal{Y}$ is equivalent to $y = Y\hat{y} = X\hat{Y}\hat{y} =: Xz$ for some unique $z \in \hat{\mathcal{Y}} := \mathcal{R}(\hat{Y}) \subseteq \mathbb{C}^\ell$. We have by (2.13)

$$
\begin{aligned}
\alpha_i &= \min_{\substack{\mathcal{Y} \subseteq \mathcal{X} \\ \dim \mathcal{Y} = i}} \max_{y \in \mathcal{Y}} \frac{y^H A y}{y^H B y} \\
&= \min_{\substack{\hat{\mathcal{Y}} \subseteq \mathbb{C}^\ell \\ \dim \hat{\mathcal{Y}} = i}} \max_{z \in \hat{\mathcal{Y}}} \frac{z^H X^H A X z}{z^H X^H B X z} = \hat{\lambda}_i,
\end{aligned}
$$

the ith eigenvalues of $X^H A X - \lambda X^H B X$. This is the first optimality of the Rayleigh-Ritz procedure.

Suppose we are seeking λ_i for $1 \le i \le k$. By Theorem 2.1, we have

$$\sum_{i=1}^{k} \lambda_i = \min_{\substack{\mathcal{R}(Y) \subseteq \mathbb{C}^n \\ Y^H BY = I_k}} \text{trace}(Y^H AY) \tag{2.14}$$

where the minimization is taken over all $Y \in \mathbb{C}^{n \times k}$ satisfying $Y^H BY = I_k$. So given $\mathcal{X} \subset \mathbb{C}^n$, the natural definition for the best approximation is to replace $\mathbb{C}^n$ by $\mathcal{X}$ to achieve

$$\min_{\substack{\mathcal{R}(Y) \subseteq \mathcal{X} \\ Y^H BY = I_k}} \text{trace}(Y^H AY). \tag{2.15}$$

Any $\mathcal{R}(Y) \subseteq \mathcal{X}$ with $Y^H BY = I_k$ can be represented uniquely by $Y = X\widehat{Y}$ for some $\widehat{Y} \in \mathbb{C}^{\ell \times k}$ such that $\widehat{Y}^H (X^H BX)\widehat{Y} = I_k$. So (2.15) becomes

$$\min_{\substack{\mathcal{R}(Y) \subseteq \mathcal{X} \\ Y^H BY = I_k}} \text{trace}(Y^H AY) = \min_{\widehat{Y}^H (X^H BX)\widehat{Y} = I_k} \text{trace}(\widehat{Y}^H (X^H BX)\widehat{Y}) = \sum_{i=1}^{k} \widehat{\lambda}_i.$$

This gives the second optimality of the Rayleigh-Ritz procedure.

The third optimality is concerned with the residual matrix

$$\mathscr{R}(A_1) := AX - BXA_1.$$

If $\mathscr{R}(A_1) = 0$, then $\mathcal{X}$ is an exact invariant subspace. So it would make sense to make $\|\mathscr{R}(A_1)\|$ as small as possible for certain matrix norm $\| \cdot \|$. The next theorem says the optimal A_1 is $X^H AX$ when X is a B-orthonormal basis matrix of $\mathcal{X}$.

Theorem 2.4. *Suppose X has B-orthonormal columns, i.e., $X^H BX = I_k$, and let $H = X^H AX$. Then for any unitarily invariant norm[1] $\| \cdot \|_{\text{ui}}$*

$$\|B^{-1/2} \mathscr{R}(H)\|_{\text{ui}} \le \|B^{-1/2} \mathscr{R}(A_1)\|_{\text{ui}} \quad \text{for all } k\text{-by-}k \ A_1. \tag{2.16}$$

2.3 Steepest descent methods

The basic idea of the steepest descent (SD) method to minimize a function value is to perform a line-search along the (opposite) direction of

---

[1] Two commonly used unitarily invariant norms are the spectral norm $\| \cdot \|_2$ and the Frobenius norm $\| \cdot \|_F$. It is natural to think of $\|B^{-1/2}(\cdot)\|_{\text{ui}}$ as the B^{-1}-unitarily invariant norm induced by the given unitarily invariant norm $\| \cdot \|_{\text{ui}}$. For example, the usual Frobenius norm can be defined by $\|C\|_F := \sqrt{\text{trace}(C^H C)}$. Correspondingly, we may define the B^{-1}-Frobenius norm by $\|C\|_{B^{-1};F} = \sqrt{\text{trace}(C^H B^{-1} C)}$.

the gradient of the function at each iteration step. Our function is $\rho(x)$ defined by (2.5) whose gradient is given by

$$\nabla \rho(x) = \frac{2}{x^{\mathrm{H}} B x} r(x),\qquad(2.17)$$

where $r(x) := Ax - \rho(x) Bx$ is the residual of $(\rho(x), x)$ as an approximate eigenpair of $A - \lambda B$. Notice that $\nabla \rho(x)$ points to the same direction as $r(x)$. Therefore, given an approximation $\boldsymbol{x}$ to u_1 and $\|\boldsymbol{x}\|_B = 1$, one step of the steepest descent method for computing (λ_1, u_1) is simply to perform a line-search:

$$\inf_{t \in \mathbb{C}} \rho(\boldsymbol{x} + t\boldsymbol{r}),\qquad(2.18)$$

where $\boldsymbol{r} = r(x)$. Since $\boldsymbol{x}^{\mathrm{H}} \boldsymbol{r} = 0$, $\boldsymbol{x}$ and $\boldsymbol{r}$ are linearly independent unless $\boldsymbol{r} = 0$ which implies $(\rho(\boldsymbol{x}), \boldsymbol{x})$ is already an exact eigenpair. An easy way to use stopping criteria is to check if

$$\frac{\|r(\boldsymbol{x})\|_2}{\|A\boldsymbol{x}\|_2 + |\rho(\boldsymbol{x})| \, \|B\boldsymbol{x}\|_2} \leq \mathtt{rtol},\qquad(2.19)$$

where $\mathtt{rtol}$ is a given relative tolerance. When it is satisfied, $(\rho(\boldsymbol{x}), \boldsymbol{x})$ will be accepted as a computed eigenpair.

We have to solve the line-search (2.18). Since such a problem arises often in the conjugate gradient methods for $A - \lambda B$, we consider the following more general line-search:

$$\inf_{t \in \mathbb{C}} \rho(x + tp),\qquad(2.20)$$

where the search direction p is the residual $r(x)$ in the SD method but will be different in the conjugate gradient method, for example. Suppose that x and p are linearly independent; otherwise $\rho(x+tp) \equiv \rho(x) = \rho(p)$. It is not difficult to show that

$$\inf_{t \in \mathbb{C}} \rho(x + tp) = \min_{|\xi|^2 + |\zeta|^2 > 0} \rho(\xi x + \zeta p).\qquad(2.21)$$

Therefore the infimum in (2.20) is the smaller eigenvalue μ of the 2×2 matrix pencil $X^{\mathrm{H}} A X - \lambda X^{\mathrm{H}} B X$, where $X = [x, p]$. Let $v = [\nu_1, \nu_2]^{\mathrm{T}}$ be the corresponding eigenvector. Then $\rho(Xv) = \mu$. Note that $Xv = \nu_1 x + \nu_2 p$. We conclude

$$\operatorname*{arginf}_{t \in \mathbb{C}} \rho(x + tp) =: t_{\mathrm{opt}} = \begin{cases} \nu_2/\nu_1, & \text{if } \nu_1 \neq 0, \\ \infty, & \text{if } \nu_1 = 0. \end{cases}\qquad(2.22)$$

Here $t_{\mathrm{opt}} = \infty$ should be interpreted in the sense of taking $t \to \infty$:

$$\lim_{t \to \infty} \rho(x + tp) = \rho(p).$$

Accordingly, we have

$$\rho(y) = \inf_{t \in \mathbb{C}} \rho(x + tp), \quad y = \begin{cases} x + t_{\text{opt}} p, & \text{if } t_{\text{opt}} \text{ is finite,} \\ p, & \text{otherwise.} \end{cases} \quad (2.23)$$

Now the simple SD method can be readily stated. We leave it to the reader.

This simple SD method can be slowly convergent in practice. This happens when the contours of $\rho(x)$ on the sphere $\{x : x^H x = 1\}$ near the eigenvector u_1 is very flat: very long stretched in one or a few directions but very short compressed in other directions. So rarely, this plain version is used in practice, but rather as a starting point for designing faster variations of the method. In what follows, we will present three ideas some or all of which can be combined to improve the method in practice. The three ideas are

- extending the search space,
- preconditioning the search direction,
- introducing block implementation.

We now explain the three ideas in detail.

Extending the search space. The key step of the SD method is the line-search (2.18) which can be interpreted as seeking the best possible approximation ρ_{new}:

$$\rho_{\text{new}} = \min_{z \in \text{span}\{x,r\}} \rho(z) \quad (2.24)$$

to λ_1 through projecting $A - \lambda B$ to the 2-dimensional subspace spanned by

$$x, \quad r = Ax - \rho Bx = (A - \rho B)x,$$

where $\rho = \rho(x)$. This subspace is in fact the 2nd order Krylov subspace $\mathcal{K}_2(A - \rho B, x)$ of $A - \rho B$ on x. Naturally, a way to accelerate the simple SD method is to use a larger Krylov subspace, i.e., the mth order Krylov subspace $\mathcal{K}_m(A - \rho B, x)$ which is spanned by

$$x, (A - \rho B)x, \ldots, (A - \rho B)^{m-1}x.$$

A better approximation to λ_1 is then obtained for $m \geq 3$ since now they are achieved by minimizing $\rho(x)$ over a larger subspace that contains $\text{span}\{x, r\} = \mathcal{K}_2(A - \rho B, x)$:

$$\rho_{\text{new}} = \min_{z \in \mathcal{K}_m(A - \rho B, x)} \rho(z). \quad (2.25)$$

This leads to the *inverse free Krylov subspace method* of Golub and Ye [12] but we will call it the *extended steepest descent method* (ESD).

Theorem 2.5 ([12]). *Suppose λ_1 is simple, i.e., $\lambda_1 < \lambda_2$, and $\lambda_1 < \rho < \lambda_2$. Let $\omega_1 < \omega_2 \leq \cdots \leq \omega_n$ be the eigenvalues of $A - \rho B$ and v_1 be an eigenvector corresponding to ω_1, and let ρ_{new} be defined by (2.25). Then*

$$\rho_{\text{new}} - \lambda_1 \leq (\rho - \lambda_1)\epsilon_m^2 + 2(\rho - \lambda_1)^{3/2}\epsilon_m \left(\frac{\|B\|_2}{\omega_2}\right)^{1/2} + O(|\rho - \lambda_1|^2), \quad (2.26)$$

where

$$\epsilon_m := \min_{f \in \mathbb{P}_{m-1}, f(\omega_1)=1} \max_{j>1} |f(\omega_j)| \leq 2\left[\Delta_\eta^{m-1} + \Delta_\eta^{-(m-1)}\right]^{-1}, \quad (2.27)$$

$\eta = \frac{\omega_2 - \omega_1}{\omega_n - \omega_1}$ *and* $\Delta_\eta = \frac{1 + \sqrt{\eta}}{1 - \sqrt{\eta}}$.

There are a few other existing results for $m = 2$ and $B = I$. Kantorovich and Akilov [21, p.617] established

$$(\rho_{\text{new}} - \lambda_1)/(\rho - \lambda_1) \lesssim \epsilon_m^2$$

for completely continuous operators. Knyazev and Skorokhodov [22] obtained something that is stronger in the sense that it is a strict inequality (i.e., without the need of ignoring high order terms). Samokish [45] presented an estimate on convergence rate for the preconditioned steepest descent method. Although his technique was for the case $B = I$, but can be made to work for the case $B \neq I$ after minor changes (see also [24, 37]). We omit stating them to limit the length of this chapter.

Preconditioning the search direction. The idea of preconditioning a linear system $Ax = b$ to $KAx = Kb$ such that KA is "almost" the identity matrix before it is iteratively solved is quite natural. After all if $KA = I$, we would have $x = Kb$ immediately. Here that KA is "almost" the identity matrix is understood either $\|KA - I\|$ is relatively small or $KA - I$ is near a low rank matrix.

But there is no such an obvious and straightforward way to precondition the eigenvalue problem $Ax = \lambda Bx$. How could any direction be more favorable than the steepest descent one when it comes to minimize $\rho(x)$? After all, we are attempting to minimize the objective function $\rho(x)$.

In what follows, we shall offer two viewpoints as to understand preconditioning an eigenvalue problem and how an effective preconditioner should be approximately constructed.

The first viewpoint is more intuitive. The rationale lies as follows. It is well known that when the contours of the objective function near its optimum are extremely elongated, at each step of the conventional steepest descent method, following the search direction which is the opposite of the gradient gets closer to the optimum on the line for a very

short while and then starts to get away because the direction doesn't
point "towards the optimum", resulting in a long zigzag path of a large
number of steps. The ideal search direction p is therefore the one such
that with its starting point at $\boldsymbol{x}$, p points to the optimum, i.e., the optimum is on the line $\{\boldsymbol{x} + tp : t \in \mathbb{C}\}$. Specifically, expand $\boldsymbol{x}$ as a linear
combination of eigenvectors u_j

$$\boldsymbol{x} = \sum_{j=1}^{n} \alpha_j u_j =: \alpha_1 u_1 + \boldsymbol{v}, \quad \boldsymbol{v} = \sum_{j=2}^{n} \alpha_j u_j. \qquad (2.28)$$

Then the ideal search direction is

$$p = \alpha u_1 + \beta \boldsymbol{v}$$

for some scalar α and $\beta \neq 0$ such that $\alpha_1 \beta - \alpha \neq 0$ (otherwise $p = \beta \boldsymbol{x}$).
Of course, this is impractical because we don't know u_1 and $\boldsymbol{v}$. But we
can construct one that is close to it. One such p is

$$p = (A - \sigma B)^{-1} \boldsymbol{r} = (A - \sigma B)^{-1}(A - \rho B)\boldsymbol{x},$$

where[2] σ is some shift near λ_1 but not equal to ρ. Let us analyze this
p. By (2.4), we find

$$p = \sum_{j=1}^{n} \mu_j \alpha_j u_j, \quad \mu_j := \frac{\lambda_j - \rho}{\lambda_j - \sigma}. \qquad (2.29)$$

Now if $\lambda_1 \leq \rho < \lambda_2$ and σ is also near λ_1 but not equal to ρ and if the
gap $\lambda_2 - \lambda_1$ is reasonably modest, then

$$\mu_j \approx 1 \quad \text{for } j > 1$$

to give a $p \approx \alpha u_1 + \boldsymbol{v}$, resulting in fast convergence. This rough but
intuitive analysis suggests that $(A - \sigma B)^{-1}$ with a suitably chosen shift
σ can be used to serve as a good preconditioner. Qualitatively, we have

Theorem 2.6. *Let $\boldsymbol{x}$ be given by (2.28), and suppose $\alpha_1 \neq 0$. If $\sigma \neq \rho$
such that*

$$\text{either } \mu_1 < \mu_j \text{ for } 2 \leq j \leq n \text{ or } \mu_1 > \mu_j \text{ for } 2 \leq j \leq n, \qquad (2.30)$$

where μ_j are defined in (2.29), then

$$\tan \theta_B(u_1, \mathcal{K}_m) \leq 2 \left[\Delta_\eta^{m-1} + \Delta_\eta^{-(m-1)} \right]^{-1} \tan \theta_B(u_1, \boldsymbol{x}), \qquad (2.31)$$

$$0 \leq \rho_{\text{new}} - \lambda_1 \leq 4 \left[\Delta_\eta^{m-1} + \Delta_\eta^{-(m-1)} \right]^{-2} \tan \theta_B(u_1, \boldsymbol{x}), \qquad (2.32)$$

---

[2] We reasonably assume also $\sigma \neq \lambda_j$ for all other j, too.

where $\mathcal{K}_m := \mathcal{K}_m([A - \sigma B]^{-1}(A - \boldsymbol{\rho}B), \boldsymbol{x})$, *and*

$$\eta = \begin{cases} \frac{\lambda_n - \sigma}{\lambda_n - \lambda_1} \cdot \frac{\lambda_2 - \lambda_1}{\lambda_2 - \sigma}, & \text{if } \mu_1 < \mu_j \text{ for } 2 \le j \le n, \\ \frac{\lambda_2 - \sigma}{\lambda_2 - \lambda_1} \cdot \frac{\lambda_n - \lambda_1}{\lambda_n - \sigma}, & \text{if } \mu_1 > \mu_j \text{ for } 2 \le j \le n, \end{cases} \qquad \Delta_\eta = \frac{1 + \sqrt{\eta}}{1 - \sqrt{\eta}}.$$

Proof. The proof is similar to the one in Saad [43] for the symmetric Lanczos method. □

The assumption (2.30) is one of the two criteria for selecting a shift σ, and the other is to make η close to 1. Three interesting cases are

- $\sigma < \lambda_1 \le \rho < \lambda_2$ under which μ_1 is smallest,
- $\lambda_1 < \sigma < \rho < \lambda_2$ under which μ_1 is biggest,
- $\lambda_1 < \rho < \sigma < \lambda_2$ under which μ_1 is smallest.

Often σ is selected as a lower bound of λ_1 as in the first case above, but it does not have to be. As for η, it is 1 for $\sigma = \lambda_1$, but since λ_1 is unknown, the best one can hope is to make $\sigma \approx \lambda_1$ through some kind of estimation.

In practice, because of high cost associated with $(A - \sigma B)^{-1}$, some forms of approximations to $(A - \sigma B)^{-1}$, such as those by incomplete decompositions LDL^H of $A - \sigma B$ or by iterative methods [9, 13, 14], such as CG, MINRES, or GMRES, are widely used.

The second viewpoint is proposed by Golub and Ye [12], based on Theorem 2.5 which reveals that the rate of convergence depend on the distribution of the eigenvalues ω_j of $A - \boldsymbol{\rho}B$, not those of the pencil $A - \lambda B$ as in the Lanczos algorithm. In particular, if all $\omega_2 = \cdots = \omega_n$, then $\epsilon_m = 0$ for $m \ge 2$ and thus

$$\boldsymbol{\rho}_{\text{new}} - \lambda_1 = O(|\boldsymbol{\rho} - \lambda_1|^2),$$

suggesting quadratic convergence. Such an extreme case, though highly welcome, is unlikely to happen in practice, but it gives us an idea that if somehow we could transform an eigenvalue problem towards such an extreme case, the transformed problem would be easier to solve. Specifically we should seek equivalent transformations that change the eigenvalues of $A - \boldsymbol{\rho}B$ as much as possible to,

one smallest isolated eigenvalue ω_1, and the rest ω_j $(2 \le j \le n)$ tightly clustered,	(2.33)

but leave those of $A - \lambda B$ unchanged. This goal is much as the one for preconditioning a linear system $Ax = b$ to $KAx = Kb$ for which a similar eigenvalue distribution for KA like (2.33) will result in swift convergence by most iterative methods.

We would like to equivalently transform the eigenvalue problem for $A - \lambda B$ to $L^{-H}(A - \lambda B)L^{-1}$ by some nonsingular L (whose inverse is easy to compute or any linear system with L is easy to solve) so that the eigenvalues of $L^{-1}(A - \rho B)L^{-H}$ distribute more or less like (2.33). Then apply one step of ESD to the pencil $L^{-1}(A - \lambda B)L^{-H}$ to find the next approximation ρ_{new}. The process repeats.

Borrowed from the incomplete decomposition idea for preconditioning a linear system, such an L can be constructed using the LDL^H decomposition of $A - \rho B$ [13, p.139] if the decomposition exists: $A - \rho B = LDL^H$, where L is lower triangular and $D = \text{diag}(\pm 1)$. Then $L^{-1}(A - \rho B)L^{-H} = D$ has the ideal eigenvalue distribution that gives $\epsilon_m = 0$ for any $m \geq 2$. Unfortunately, this simple solution is impractical in practice for the following reasons:

1. The decomposition may not exist at all. In theory, the decomposition exists if all of the leading principle submatrices of $A - \rho B$ are nonsingular.

2. If the decomposition does exist, it may not be numerically stable to compute, especially when ρ comes closer and closer to λ_1.

3. The sparsity in A and B is most likely destroyed, leaving L significantly denser than A and B combined. This makes all ensuing computations much more expensive.

A more practical solution is, however, through an incomplete LU factorization (see [44, Chapter 10]), to get

$$A - \rho B \approx LDL^H,$$

where "$\approx$" includes not only the usual "approximately equal", but also the case when $(A - \rho B) - LDL^H$ is approximately a low rank matrix, and $D = \text{diag}(\pm 1)$. Such an L changes from one step of the algorithm to another. In practice, often we may use one fixed preconditioner for all or several iterative steps. Using a constant preconditioner is certainly not optimal: it likely won't give the best rate of convergence per step and thus increases the number of total iterative steps but it can reduce overall cost because it saves work in preconditioner constructions and thus reduces cost per step. The basic idea of using a step-independent preconditioner is to find a σ that is close to λ_1, and perform an incomplete LDL^H decomposition of

$$A - \sigma B \approx LDL^H$$

and transform $A - \lambda B$ accordingly before applying SD or ESD. Now the rate of convergence is determined by the eigenvalues of

$$\widehat{C} = L^{-1}(A - \sigma B)L^{-H} + (\sigma - \rho)L^{-1}BL^{-H} \approx D$$

which would have a better spectral distribution so long as $(\sigma-\rho)L^{-1}BL^{-H}$ is small relative to $\widehat{C}$. When $\sigma < \lambda_1$, $A - \sigma B \succ 0$ and the incomplete LDL^H factorization becomes incomplete Cholesky factorization.

We have insisted so far about applying SD or ESD straightforwardly to the transformed problem. There is another way, perhaps, better: only symbolically applying SD or ESD to the transformed problem as a derivation stage for a preconditioned method that always projects the original pencil $A - \lambda B$ directly every step. The only difference is now the projecting subspaces are preconditioned.

Suppose $A - \lambda B$ is transformed to $\widehat{A} - \lambda \widehat{B} := L^{-1}(A - \lambda B)L^{-H}$. Consider a typical step of ESD applied to $\widehat{A} - \lambda \widehat{B}$. For the purpose of distinguishing notational symbols, we will put hats on all those for $\widehat{A} - \lambda \widehat{B}$. The typical step of ESD is

$$\boxed{\begin{array}{l} \text{compute the smallest eigenvalue } \mu \text{ and corresponding eigen-} \\ \text{vector } v \text{ of } \widehat{Z}^H(\widehat{A}-\lambda\widehat{B})\widehat{Z}, \text{ where } \widehat{Z} \in \mathbb{C}^{n\times m} \text{ is a basis matrix} \\ \text{of Krylov subspace } \mathcal{K}_m(\widehat{A} - \hat{\rho}\widehat{B}, \hat{\boldsymbol{x}}). \end{array}} \quad (2.34)$$

Notice $\left[\widehat{A} - \hat{\rho}\widehat{B}\right]^j \hat{\boldsymbol{x}} = L^H\left[(LL^H)^{-1}(A - \hat{\rho}B)\right]^j (L^{-H}\hat{\boldsymbol{x}})$ to see

$$L^{-H} \cdot \mathcal{K}_m(\widehat{A} - \hat{\rho}\widehat{B}, \hat{\boldsymbol{x}}) = \mathcal{K}_m(K(A - \hat{\rho}B), \boldsymbol{x}),$$

where $\boldsymbol{x} = L^{-H}\hat{\boldsymbol{x}}$ and $K = (LL^H)^{-1}$. So $Z = L^{-H}\widehat{Z}$ is a basis matrix of Krylov subspace $\mathcal{K}_m(K(A - \hat{\rho}B), \boldsymbol{x})$. Since also

$$\widehat{Z}^H(\widehat{A} - \lambda\widehat{B})\widehat{Z} = (L^{-H}\widehat{Z})^H(A - \lambda B)(L^{-H}\widehat{Z}),$$

$$\hat{\rho} = \frac{\hat{\boldsymbol{x}}^H\widehat{A}\hat{\boldsymbol{x}}}{\hat{\boldsymbol{x}}^H\widehat{B}\hat{\boldsymbol{x}}} = \frac{\boldsymbol{x}^HA\boldsymbol{x}}{\boldsymbol{x}^HB\boldsymbol{x}} = \rho,$$

the typical step (2.34) can be reformulated equivalently to

$$\boxed{\begin{array}{l} \text{compute the smallest eigenvalue } \mu \text{ and corresponding eigen-} \\ \text{vector } v \text{ of } Z^H(A - \lambda B)Z, \text{ where } Z \in \mathbb{C}^{n\times m} \text{ is a ba-} \\ \text{sis matrix of Krylov subspace } \mathcal{K}_m(K(A - \rho B), \boldsymbol{x}), \text{ where} \\ K = (LL^H)^{-1}. \end{array}} \quad (2.35)$$

Introducing block implementation. The convergence rate of ESD with a preconditioner $K \succ 0$ is determined by the eigenvalues $\omega_1 < \omega_2 \leq \cdots \leq \omega_n$ of $K^{1/2}(A - \rho B)K^{1/2}$, and it can still be very slow if λ_2 is very close to λ_1 relative to λ_n in which case $\omega_1 \approx \omega_2$.

Often in practice, there are needs to compute the first few eigenpairs, not just the first one. For that purpose, block variations of the methods become particularly attractive for at least the following reasons:

1. they can simultaneously compute the first k eigenpairs (λ_j, u_j);

2. they run more efficiently on modern computer architecture because more computations can be organized into matrix-matrix multiplication type;

3. they have better rates of convergence to the desired eigenpairs and save overall cost by using a block size that is slightly bigger than the number of asked eigenpairs.

In summary, the benefits of using a block variation are similar to those of using the simultaneous subspace iteration *vs.* the power method [46].

A block variation starts with $X \in \mathbb{C}^{n \times n_b}$ with $\text{rank}(X) = n_b$, instead of just one vector $x \in \mathbb{C}^n$ previously for the single-vector steepest descent methods. Here either the jth column of X is already an approximation to u_j or the subspace $\mathcal{R}(X)$ is a good approximation to the generalized invariant subspace spanned by u_j for $1 \leq j \leq n_b$ or the canonical angles from $\mathcal{R}([u_1, \ldots, u_k])$ to $\mathcal{R}(X)$ are nontrivial, where $k \leq n_b$ is the number of desired eigenpairs. In the latter two cases, a preprocessing is needed to turn the case into the first case:

1. solve the eigenvalue problem $X^H(A - \lambda B)X$ to get $(X^H A X) W = (X^H B X) W \Omega$, where $\Omega = \text{diag}(\rho_1, \rho_2, \ldots, \rho_{n_b})$ is the diagonal matrix of eigenvalues in ascending order, and W is the eigenvector matrix;

2. reset $X := XW$.

So we will assume henceforth the jth column of the given X is an approximation to u_j. Now consider generalizing the steepest descent method to a block one. Its typical iterative step may well look like the following. Let

$$X = [x_1, x_2, \ldots, x_{n_b}] \in \mathbb{C}^{n \times n_b}$$

whose jth column x_j approximates u_j and

$$\Omega = \text{diag}(\rho_1, \rho_2, \ldots, \rho_{n_b})$$

whose jth diagonal entry $\rho_j = \rho(x_j)$ approximates λ_j. We may well assume X has B-orthonormal columns, i.e., $X^H B X = I$. Define the residual matrix

$$R = [r(x_1), r(x_2), \ldots, r(x_{n_b})] = AX - BX\Omega.$$

The key iterative step of the block steepest descent (BSD) method for computing the next set of approximations is as follows:

1. compute a basis matrix Z of $\mathcal{R}([X, R])$ by, e.g., MGS in the B-inner product, keeping in mind that X already has B-orthonormal columns;

2. find the first n_b eigenpairs of $Z^H A Z - \lambda Z^H B Z$ by, e.g., one of LAPACK's subroutines [1, p.25] because of its small scale, to get $(Z^H A Z)W = (Z^H B Z)W\Omega_{\text{new}}$, where

$$\Omega_{\text{new}} = \text{diag}(\boldsymbol{\rho}_{\text{new};1}, \boldsymbol{\rho}_{\text{new};2}, \ldots, \boldsymbol{\rho}_{\text{new};n_b});$$

3. set $\boldsymbol{X}_{\text{new}} = ZW$.

This is in fact the stronger version of **Simultaneous Rayleigh Quotient Minimization Method**, called SIRQIT-G2, in Longsine and Mc-Cormick [30]. To introduce the block extended steepest descent (BESD) method, we notice that $r(\boldsymbol{x}_j) = (A - \boldsymbol{\rho}_j B)\boldsymbol{x}_j$ and thus

$$\mathcal{R}([\boldsymbol{X},\boldsymbol{R}]) = \sum_{j=1}^{n_b} \mathcal{R}([\boldsymbol{x}_j, (A - \boldsymbol{\rho}_j B)\boldsymbol{x}_j])$$

$$= \sum_{j=1}^{n_b} \mathcal{K}_2(A - \boldsymbol{\rho}_j B, \boldsymbol{x}_j).$$

BESD is simply to extend each Krylov subspace $\mathcal{K}_2(A - \boldsymbol{\rho}_j B, \boldsymbol{x}_j)$ to a high order one, and of course different Krylov subspaces can be expanded to different orders. For simplicity, we will expand each to the mth order. The new extended search subspace is now

$$\sum_{j=1}^{n_b} \mathcal{K}_m(A - \boldsymbol{\rho}_j B, \boldsymbol{x}_j). \tag{2.36}$$

Define the linear operator

$$\mathscr{R} : X \in \mathbb{C}^{n \times n_b} \to \mathscr{R}(X) = AX - BX\Omega \in \mathbb{C}^{n \times n_b}.$$

Then the subspace in (2.36) can be compactly written as

$$\mathcal{K}_m(\mathscr{R}, X) = \text{span}\{\boldsymbol{X}, \mathscr{R}(\boldsymbol{X}), \ldots, \mathscr{R}^{m-1}(\boldsymbol{X})\}, \tag{2.37}$$

where $\mathscr{R}^i(\,\cdot\,)$ is understood as successively applying the operator $\mathscr{R}$ i times, e.g., $\mathscr{R}^2(X) = \mathscr{R}(\mathscr{R}(X))$.

As to incorporate suitable preconditioners, in light of our extensive discussions before, the search subspace should be modified to

$$\sum_{j=1}^{n_b} \mathcal{K}_m(K_j(A - \boldsymbol{\rho}_j B), \boldsymbol{x}_j), \tag{2.38}$$

where K_j are the preconditioners, one for each approximate eigenpair $(\boldsymbol{\rho}_j, \boldsymbol{x}_j)$ for $1 \le j \le n_b$. As before, K_j can be constructed in one of the following two ways:

- K_j is an approximate inverse of $A - \tilde{\rho}_j B$ for some $\tilde{\rho}_j$ different from ρ_j, ideally closer to λ_j than to any other eigenvalue of $A - \lambda B$. But this requirement on $\tilde{\rho}_j$ is impractical because the eigenvalues of $A - \lambda B$ are unknown. A compromise would be to make $\tilde{\rho}_j$ close but not equal to ρ_j than to any other ρ_j.

- Perform an incomplete LDL^H factorization (see [44, Chapter 10]) $A - \rho_j B \approx L_j D_j L_j^H$, where "$\approx$" includes not only the usual "approximately equal", but also the case when $(A - \rho_j B) - L_j D_j L_j^H$ is approximately a low rank matrix, and $D_j = \text{diag}(\pm 1)$. Finally set $K_j = L_j L_j^H$.

Algorithm 2.2 is the general framework of a Block Preconditioned Extended Steepest Descent method (BPESD) which embeds many steepest descent methods into one. In particular,

1. With $n_b = 1$, it gives various single-vector steepest descent methods:

   - Steepest Descent method (SD): $m = 2$ and all preconditioners $K_{\ell;j} = I$;
   - Preconditioned Steepest Descent method (PSD): $m = 2$;
   - Extended Steepest Descent method (ESD): all preconditioners $K_{\ell;j} = I$;
   - Preconditioned Extended Steepest Descent method (PESD).

2. With $n_b > 1$, various block steepest descent methods are born:

   - Block Steepest Descent method (BSD): $m = 2$ and all preconditioners $K_{\ell;j} = I$;
   - Block Preconditioned Steepest Descent method (BPSD): $m = 2$;
   - Block Extended Steepest Descent method (BESD): all preconditioners $K_{\ell;j} = I$;
   - Block Preconditioned Extended Steepest Descent method (BPESD).

This framework is essentially the one implied in [40, section 4].

There are four important implementation issues to worry about in turning Algorithm 2.2 into a piece of working code.

1. In (2.38), a different preconditioner is used for each and every approximate eigenpair $(\rho_{\ell;j}, x_{\ell;j})$ for $1 \le j \le n_b$. While, conceivably, doing so will speed up convergence for each approximate eigenpair because each preconditioner can be constructed to make that approximate eigenpair converge faster, but the cost in constructing these preconditioners may likely be too heavy to bear. A more practical approach would be to use

Algorithm 2.2 Extended block preconditioned steepest descent method

Given an initial approximation $X_0 \in \mathbb{C}^{n \times n_b}$ with $\mathrm{rank}(X_0) = n_b$, and an integer $m \geq 2$, the algorithm attempts to compute approximate eigenpairs to (λ_j, u_j) for $1 \leq j \leq n_b$.

1: compute the eigen-decomposition: $(X_0^{\mathrm{H}} A X_0)W = (X_0^{\mathrm{H}} B X_0)W\Omega_0$, where $W^{\mathrm{H}}(X_0^{\mathrm{H}} B X_0)W = I$, $\Omega_0 = \mathrm{diag}(\rho_{0;1}, \rho_{0;2}, \ldots, \rho_{0;n_b})$;

2: $X_0 \equiv [x_{0;1}, x_{0;2}, \ldots, x_{0;n_b}] = X_0 W$;

3: **for** $\ell = 0, 1, \ldots$ **do**

4:    test convergence and lock up the converged (detail to come later);

5:    construct preconditioners $K_{\ell;j}$ for $1 \leq j \leq n_b$;

6:    compute a basis matrix Z of the subspace (2.38) with $\boldsymbol{\rho}_j = \rho_{\ell;j}$ and $\boldsymbol{x}_j = x_{\ell+1;j}$;

7:    compute the n_b smallest eigenvalues and corresponding eigenvectors of $Z^{\mathrm{H}}(A - \lambda B)Z$ to get $(Z^{\mathrm{H}} A Z)W = (Z^{\mathrm{H}} B Z)W\Omega_\ell$, where $W^{\mathrm{H}}(Z^{\mathrm{H}} B Z)W = I$, $\Omega_{\ell+1} = \mathrm{diag}(\rho_{\ell+1;1}, \rho_{\ell+1;2}, \ldots, \rho_{\ell+1;n_b})$;

8:    $X_{\ell+1} \equiv [x_{\ell+1;1}, x_{\ell+1;2}, \ldots, x_{\ell+1;n_b}] = ZW$;

9: **end for**

10: **return** approximate eigenpairs to (λ_j, u_j) for $1 \leq j \leq n_b$.

one preconditioner K_ℓ for all $K_{\ell;j}$ aiming at speeding up the convergence of $(\rho_{\ell;1}, x_{\ell;1})$ (or the first few approximate eigenpairs for tightly clustered eigenvalues). Once it (or the first few in the case of a tight cluster) is determined to be sufficiently accurate, the converged eigenpairs are locked up and deflated and a new preconditioner is computed to aim at the next non-converged eigenpairs, and the process continues. We will come back to discuss the deflation issue, i.e., Line 4 of Algorithm 2.2.

2. Consider implementing Line 6, i.e., generating a basis matrix for the subspace (2.38). In the most general case, Z can be obtained by packing the basis matrices of all

$$\mathcal{K}_m(K_{\ell;j}(A - \rho_{\ell;j}B), x_{\ell;j}) \quad \text{for } 1 \leq j \leq n_b$$

together. There could be two problems with this: 1) such Z could be ill-conditioned, i.e., the columns of Z may not be sufficiently numerically linearly independent, and 2) the arithmetic operations in building a basis for each $\mathcal{K}_m(K_{\ell;j}(A - \rho_{\ell;j}B), x_{\ell;j})$ are mostly matrix-vector multiplications, straying from one of the purposes: performing most arithmetic operations through matrix-matrix multiplications in order to achieve high performance on modern computers. To address these two problems, we do a tradeoff of using $K_{\ell;j} \equiv K_\ell$ for all j. This may likely degrade the effectiveness of the preconditioner per step in terms of rate of convergence for all approximate eigenpairs $(\rho_{\ell;j}, x_{\ell;j})$ but may achieve overall gain in using less time because then the code will run much faster in matrix-

matrix operations, not to mention the saving in constructing just one
preconditioner K_ℓ instead of n_b different ones. To simplify our discussion below, we will drop the subscript ℓ for readability. Since $K_{\ell;j} \equiv K$
for all j, (2.38) is the same as

$$\mathcal{K}_m(K\mathcal{R}, X) = \mathrm{span}\{X, K\mathcal{R}(X), \ldots, [K\mathcal{R}]^{m-1}(X)\}, \qquad (2.39)$$

where $[K\mathcal{R}]^i(\,\cdot\,)$ is understood as successively applying the operator $K\mathcal{R}$
i times, e.g., $[K\mathcal{R}]^2(X) = K\mathcal{R}_\ell(K\mathcal{R}(X))$. A basis matrix

$$Z = [Z_1, Z_2, \ldots, Z_m]$$

can be computed by the following block Arnoldi-like process in the B-
inner product [40, Algorithm 5].

> 1: $Z_1 = X$ (recall $X^H BX = I_{n_b}$ already);
> 2: **for** $i = 2$ to m **do**
> 3: $Y = K(AZ_{i-1} - B\Omega Z_{i-1})$;
> 4:   **for** $j = 1$ to $i-1$ **do**
> 5: $T = Z_j^H BY$, $Y = Y - Z_j T$;
> 6:   **end for**
> 7: $Z_i T = Y$ (MGS in the B-inner product);
> 8: **end for**

$$(2.40)$$

There is a possibility that at Line 7 of (2.40), Y is numerically not of
full column rank. If it happens, it poses no difficulty at all. In running
MGS on Y's columns, anytime if a column is deemed linearly dependent
on previous columns, that column should be deleted, along with the
corresponding ρ_j from Ω to shrink its size by 1 as well. At the completion
of MGS, Z_i will have fewer columns than Y and the size of Ω is shrunk
accordingly. Finally, at the end, the columns of Z are B-orthonormal,
i.e., $Z^H BZ = I$ (of apt size) which may fail to an unacceptably level due
to roundoff; so some form of re-orthogonalization should be incorporated.

3. At Line 4, a test for convergence is required. The same criteria
(2.19) can be used: $(\rho_{\ell;j}, x_{\ell;j})$ is considered acceptable if

$$\frac{\|r_{\ell;j}\|_2}{\|Ax_{\ell;j}\|_2 + |\rho_{\ell;j}|\,\|Bx_{\ell;j}\|_2} \leq \mathtt{rtol}$$

where $\mathtt{rtol}$ is a pre-set relative tolerance. Usually the eigenvalues λ_j
are converged to in order, i.e., the smallest eigenvalues emerge first. All
acceptable approximate eigenpairs should be locked in, say, a $k_{\mathrm{cvgd}} \times
k_{\mathrm{cvgd}}$ diagonal matrix[3] D for converged eigenvalues and an $n \times k_{\mathrm{cvgd}}$
tall matrix U for eigenvectors such that

$$AU \approx BUD, \quad U^H BU \approx I$$

---

[3] In actual programming code, it is likely an 1-D array. But we use a diagonal
matrix for the sake of presentation.

to an acceptable level of accuracy. Every time a converged eigenpair is detected, delete the converged $\rho_{\ell;j}$ and $x_{\ell;j}$ from Ω_ℓ and X_ℓ, respectively, and expand D and U to lock up the pair, accordingly. At the same time, either reduce n_b by 1 or append a (possibly random) B-orthogonal column to X to maintain n_b unchanged. There are two different ways to avoid recomputing any of the converged eigenpairs – a process called **deflation**.

1. At Line 7 in the above block Arnoldi-like process (2.40), each column of Z_{j+1} is B-orthogonalized against U.

2. Modify $A - \lambda B$ in form, but not explicitly, to $(A + \zeta BUU^{\mathrm{H}}B) - \lambda B$, where ζ is a real number intended to move λ_j for $1 \le j \le k_{\mathrm{cvgd}}$ to $\lambda_j + \zeta$; so it should be selected such that $\zeta + \lambda_1 \ge \lambda_{k_{\mathrm{cvgd}}+n_b+1}$.

But if there is a good way to pick a ζ such that $\zeta + \lambda_1 \ge \lambda_{k_{\mathrm{cvgd}}+n_b+1}$, the second approach is easier to use in implementation than the first one for which, if not carefully implemented, rounding errors can make $\mathcal{R}(Z)$ crawl into $\mathcal{R}(U)$ unnoticed.

2.4   Locally optimal CG methods

As is well-known, the slow convergence of the plain steepest descent method is due to the extreme flat contours of the objective function near (sometimes local) optimal points. The nonlinear conjugate gradient method is another way, besides preconditioning technique, to move the searching direction away from the steepest descent direction. Originally, the conjugate gradient (CG) method was invented in 1950s by Hestenes and Stiefel [17] for solving linear system $Hx = b$ with Hermitian and positive definite H, and later was interpreted as an iterative method for large scale linear systems. This is so-called the *linear CG* method [9, 13, 35]. In the 1960s, it was extended by Fletcher and Reeves [11] as an iterative method for solving nonlinear optimization problems (see also [35, 48]). We shall call the resulting method the *nonlinear CG* method. Often we leave out the word "linear" and "nonlinear" and simply call either method the CG method when no confusion can arise from this.

Because of the optimality properties (2.8) of the Rayleigh quotient $\rho(x)$, it is natural to apply the nonlinear CG method to compute the first eigenpair and, with the aid of deflation, the first few eigenpairs of $A - \lambda B$. The article [7] by Bradbury and Fletcher seems to be the first one to do just that.

However, it is suggested [23] that the locally optimal CG (LOCG) method [39, 49] is more suitable for the symmetric eigenvalue problem. In its simplest form, LOCG for our eigenvalue problem $A - \lambda B$ is obtained by simply modifying the line-search (2.24) for the SD method to

$$\rho_{\text{new}} = \min_{x \in \text{span}\{\boldsymbol{x}, \boldsymbol{x}_{\text{old}}, \boldsymbol{r}\}} \rho(x), \tag{2.41}$$

where $\boldsymbol{x}_{\text{old}}$ is the approximate eigenvector to u_1 from the previous iterative step.

The three ideas we explained in the previous subsection to improve the plain SD method can be introduced to improve the approximation given by (2.41), too, upon noticing the search space in (2.41) is

$$\mathcal{K}_2(A - \rho B, \boldsymbol{x}) + \mathcal{R}(\boldsymbol{x}_{\text{old}}),$$

making it possible for us to 1) extend the search space, 2) precondition the search direction $\boldsymbol{r}$, and 3) introduce block implementation, in the same way as we did for the plain SD method.

All things considered, we now present an algorithmic framework: Algorithm 2.3, *Locally Optimal Block Preconditioned Extended Conjugate Gradient method* (LOBPECG) which has implementation choices:

Algorithm 2.3 Locally optimal block preconditioned extended conjugate gate gradient method (LOBPECG)

Given an initial approximation $X_0 \in \mathbb{C}^{n \times n_b}$ with $\text{rank}(X_0) = n_b$, and an integer $m \geq 2$, the algorithm attempts to compute approximate eigenpairs to (λ_j, u_j) for $1 \leq j \leq n_b$.

1: compute the eigen-decomposition: $(X_0^{\text{H}} A X_0) W = (X_0^{\text{H}} B X_0) W \Omega_0$,
   where $W^{\text{H}}(X_0^{\text{H}} B X_0) W = I$, $\Omega_0 = \text{diag}(\rho_{0;1}, \rho_{0;2}, \ldots, \rho_{0;n_b})$;
2: $X_0 \equiv [x_{0;1}, x_{0;2}, \ldots, x_{0;n_b}] = X_0 W$, $X_{-1} = 0$;
3: **for** $\ell = 0, 1, \ldots$ **do**
4:   test convergence and lock up the converged;
5:   construct preconditioners $K_{\ell;j}$ for $1 \leq j \leq n_b$;
6:   compute a basis matrix Z of the subspace

$$\sum_{j=1}^{n_b} \mathcal{K}_m(K_{\ell;j}(A - \rho_{\ell;j} B), x_{\ell;j}) + \mathcal{R}(X_{\ell-1}); \tag{2.42}$$

7:   compute the n_b smallest eigenvalues and the corresponding eigenvectors of $Z^{\text{H}}(A - \lambda B)Z$ to get $(Z^{\text{H}} A Z)W = (Z^{\text{H}} B Z)W\Omega_\ell$, where $W^{\text{H}}(Z^{\text{H}} B Z)W = I$, $\Omega_{\ell+1} = \text{diag}(\rho_{\ell+1;1}, \rho_{\ell+1;2}, \ldots, \rho_{\ell+1;n_b})$;
8:   $X_{\ell+1} \equiv [x_{\ell+1;1}, x_{\ell+1;2}, \ldots, x_{\ell+1;n_b}] = ZW$;
9: **end for**
10: **return** approximate eigenpairs to (λ_j, u_j) for $1 \leq j \leq n_b$.

- block size n_b;
- preconditioners varying with iterative steps, with approximate eigenpairs, or not;
- the dimension m of Krylov subspaces in extending the search subspace at each iterative step. It may also vary with iterative steps, too.

The four important implementation issues we discussed for Algorithm 2.2 (BPESD) after its introduction essentially apply here, except some changes are needed in the computation of Z at Line 6 of Algorithm 2.3.

First $X_{\ell-1}$ can be replaced by something else while the subspace (2.42) remains the same. Specifically, we modify Lines 2, 6, and 8 of Algorithm 2.3 to

---

2:    $X_0 \equiv [x_{0;1}, x_{0;2}, \dots, x_{0;n_b}] = X_0 W$, and $Y_0 = 0$;

6:        compute a basis matrix Z of the subspace

$$\sum_{j=1}^{n_b} \mathcal{K}_m(K_{\ell;j}(A - \rho_{\ell;j}B), x_{\ell;j}) + \mathcal{R}(Y_\ell) \qquad (2.43)$$

such that $\mathcal{R}(Z_{(:,1:n_b)}) = \mathcal{R}(X_\ell)$, and let n_Z be the number of the columns of Z;

8:    $X_{\ell+1} \equiv [x_{\ell+1;1}, x_{\ell+1;2}, \dots, x_{\ell+1;n_b}] = ZW$,
$Y_{\ell+1} = Z_{(:,n_b+1:n_Z)}W_{(n_b+1:n_Z,:)}$;

This idea is basically the same as the one in [18, 23]. Next we will compute a basis matrix for the subspace (2.43) (or (2.42)). For better performance (by using more matrix-matrix multiplications), we will assume $K_{\ell;j} \equiv K_\ell$ for all j for simplification. Dropping the subscript ℓ for readability, we see (2.43) is the same as

$$\mathcal{K}_m(K\mathcal{R}, X) + \mathcal{R}(Y) = \text{span}\{X, K\mathcal{R}(X), \dots, [K\mathcal{R}]^{m-1}(X)\} + \mathcal{R}(Y).$$
$$(2.44)$$

We will first compute a basis matrix $[Z_1, Z_2, \dots, Z_m]$ for $\mathcal{K}_m(K\mathcal{R}, X)$ by the Block Arnoldi-like process in the B-inner product (2.40). In particular, $Z_1 = X$. Then B-orthogonalize Y against $[Z_1, Z_2, \dots, Z_m]$ to get Z_{m+1} satisfying $Z_{m+1}^H B Z_{m+1} = I$. Finally take $Z = [Z_1, Z_2, \dots, Z_{m+1}]$.

3   Min-max principles for a positive semidefinite pencil

Let $A - \lambda B$ be an $n \times n$ positive semidefinite pencil, i.e., A and B are Hermitian and there is a real scalar λ_0 such that $A - \lambda_0 B$ is positive semidefinite. Note that this does not demand anything on the regularity of $A - \lambda B$, i.e., a positive semidefinite matrix pencil can be either regular (meaning $\det(A - \lambda B) \not\equiv 0$) or singular (meaning $\det(A - \lambda B) \equiv 0$ for all $\lambda \in \mathbb{C}$).

Let the integer triplet (n_-, n_0, n_+) be the *inertia* of B, meaning B has n_- negative, n_0 0, and n_+ positive eigenvalues, respectively. Necessarily

$$r := \operatorname{rank}(B) = n_+ + n_-. \tag{3.1}$$

We say $\mu \neq \infty$ is a *finite eigenvalue* of $A - \lambda B$ if

$$\operatorname{rank}(A - \mu B) < \max_{\lambda \in \mathbb{C}} \operatorname{rank}(A - \lambda B), \tag{3.2}$$

and $x \in \mathbb{C}^n$ is a corresponding *eigenvector* if $0 \neq x \notin \mathcal{N}(A) \cap \mathcal{N}(B)$ satisfies

$$A x = \mu B x, \tag{3.3}$$

or equivalently, $0 \neq x \in \mathcal{N}(A - \mu B)\backslash(\mathcal{N}(A) \cap \mathcal{N}(B))$. Let k_+ and k_- be two nonnegative integers such that $k_+ \leq n_+$, $k_- \leq n_-$, and $k_+ + k_- \geq 1$, and set

$$J_k = \begin{bmatrix} I_{k_+} & \\ & -I_{k_-} \end{bmatrix} \in \mathbb{C}^{k \times k}, \quad k = k_+ + k_-. \tag{3.4}$$

Theorem 3.1 ([29]). *If $A - \lambda B$ is positive semidefinite, then $A - \lambda B$ has $r = \operatorname{rank}(B)$ finite eigenvalues all of which are real.*

In what follows, if $A - \lambda B$ is positive semidefinite, we will denote its finite eigenvalues by $\lambda_i^{\pm}$ arranged in the order:

$$\lambda_{n_-}^- \leq \cdots \leq \lambda_1^- \leq \lambda_1^+ \leq \cdots \leq \lambda_{n_+}^+. \tag{3.5}$$

For the case of a regular Hermitian pencil $A - \lambda B$ (i.e., $\det(A - \lambda B) \not\equiv 0$), Theorem 3.2 is a special case of the ones considered in [6, 34]. For a diagonalizable positive semidefinite Hermitian pencil $A - \lambda B$ with nonsingular B, Theorem 3.2 was implied in [26, 53]. Recall that a positive semidefinite Hermitian pencil $A - \lambda B$ can possibly be a singular pencil; so the condition of Theorem 3.2 does not exclude a singular pencil $A - \lambda B$ which was not considered before [29], not to mention that B may possibly be singular.

Theorem 3.2. *Let $A - \lambda B$ be a positive semidefinite Hermitian pencil. Then for $1 \leq i \leq n_+$*

$$\lambda_i^+ = \sup_{\substack{\mathcal{X} \\ \operatorname{codim} \mathcal{X} = i-1}} \inf_{\substack{x \in \mathcal{X} \\ x^H B x = 1}} x^H A x = \sup_{\substack{\mathcal{X} \\ \operatorname{codim} \mathcal{X} = i-1}} \inf_{\substack{x \in \mathcal{X} \\ x^H B x > 0}} \frac{x^H A x}{x^H B x}, \tag{3.6a}$$

$$\lambda_i^+ = \inf_{\substack{\mathcal{X} \\ \dim \mathcal{X} = i}} \sup_{\substack{x \in \mathcal{X} \\ x^H B x = 1}} x^H A x = \inf_{\substack{\mathcal{X} \\ \dim \mathcal{X} = i}} \sup_{\substack{x \in \mathcal{X} \\ x^H B x > 0}} \frac{x^H A x}{x^H B x}, \tag{3.6b}$$

and for $1 \le i \le n_-$,

$$\lambda_i^- = - \sup_{\substack{\mathcal{X} \\ \text{codim}\,\mathcal{X}=i-1}} \inf_{x^H Bx=-1} x^H Ax = \inf_{\substack{\mathcal{X} \\ \text{codim}\,\mathcal{X}=i-1}} \sup_{\substack{x\in\mathcal{X} \\ x^H Bx<0}} \frac{x^H Ax}{x^H Bx}, \quad (3.6c)$$

$$\lambda_i^- = - \inf_{\substack{\mathcal{X} \\ \dim\mathcal{X}=i}} \sup_{\substack{x\in\mathcal{X} \\ x^H Bx=-1}} x^H Ax = \sup_{\substack{\mathcal{X} \\ \dim\mathcal{X}=i}} \inf_{\substack{x\in\mathcal{X} \\ x^H Bx<0}} \frac{x^H Ax}{x^H Bx}. \quad (3.6d)$$

In particular, setting $i = 1$ *in* (3.6) *gives*

$$\lambda_1^+ = \inf_{x^H Bx>0} \frac{x^H Ax}{x^H Bx}, \quad \lambda_1^- = \sup_{x^H Bx<0} \frac{x^H Ax}{x^H Bx}. \quad (3.7)$$

All "inf" and "sup" can be replaced by "min" and "max" if $A - \lambda B$ *is positive definite or positive semidefinite but diagonalizable* [4].

The following theorem for the case when B is also nonsingular is due to Kovač-Striko and Veselić [25]. But in this general form, it is due to [29].

Theorem 3.3 ([29]). *Let* $A - \lambda B$ *be a Hermitian pencil of order* n.

1. *Suppose* $A - \lambda B$ *is positive semidefinite. Let* $X \in \mathbb{C}^{n\times k}$ *satisfying* $X^H BX = J_k$, *and denote by* $\mu_i^{\pm}$ *the eigenvalues of* $X^H AX - \lambda X^H BX$ *arranged in the order:*

$$\mu_{k_-}^- \le \cdots \le \mu_1^- \le \mu_1^+ \le \cdots \le \mu_{k_+}^+. \quad (3.8)$$

Then

$$\lambda_i^+ \le \mu_i^+ \le \lambda_{i+n-k}^+, \quad \text{for } 1 \le i \le k_+, \quad (3.9)$$
$$\lambda_{j+n-k}^- \le \mu_i^- \le \lambda_i^-, \quad \text{for } 1 \le j \le k_-, \quad (3.10)$$

where we set $\lambda_i^+ = \infty$ *for* $i > n_+$ *and* $\lambda_j^- = -\infty$ *for* $j > n_-$.

2. *If* $A - \lambda B$ *is positive semidefinite, then*

$$\inf_{X^H BX=J_k} \text{trace}(X^H AX) = \sum_{i=1}^{k_+} \lambda_i^+ - \sum_{i=1}^{k_-} \lambda_i^-. \quad (3.11)$$

   (a) *The infimum is attainable, if there exists a matrix* $X_{\min}$ *that satisfies* $X_{\min}^H BX_{\min} = J_k$ *and whose first* k_+ *columns consist of the eigenvectors associated with the eigenvalues* λ_j^+ *for* $1 \le j \le k_+$ *and whose last* k_- *columns consist of the eigenvectors associated with the eigenvalues* λ_i^- *for* $1 \le i \le k_-$.

---

[4] Hermitian pencil $A - \lambda B$ is *diagonalizable* if there exists a nonsingular matrix W such that both $W^H AW$ and $W^H BW$ are diagonal.

(b) If $A - \lambda B$ is positive definite or positive semidefinite but diagonalizable, then the infimum is attainable.

(c) When the infimum is attained by $X_{\min}$, there is a Hermitian $A_0 \in \mathbb{C}^{k \times k}$ whose eigenvalues are $\lambda_i^{\pm}$, $i = 1, 2, \ldots, k_{\pm}$ such that

$$X_{\min}^{\mathrm{H}} B X_{\min} = J_k, \quad A X_{\min} = B X_{\min} A_0.$$

3. $A - \lambda B$ is a positive semidefinite pencil if and only if

$$\inf_{X^{\mathrm{H}} B X = J_k} \operatorname{trace}(X^{\mathrm{H}} A X) > -\infty. \tag{3.12}$$

4. If $\operatorname{trace}(X^{\mathrm{H}} A X)$ as a function of X subject to $X^{\mathrm{H}} B X = J_k$ has a local minimum, then $A - \lambda B$ is a positive semidefinite pencil and the minimum is global.

4   Linear response eigenvalue problem

We are interested in solving the standard eigenvalue problem of the form:

$$\begin{bmatrix} 0 & K \\ M & 0 \end{bmatrix} \begin{bmatrix} y \\ x \end{bmatrix} = \lambda \begin{bmatrix} y \\ x \end{bmatrix}, \tag{4.1}$$

where K and M are $n \times n$ real symmetric positive semidefinite matrices and one of them is definite. We refer to it as a *linear response (LR) eigenvalue problem* because it is equivalent to the original LR eigenvalue problem

$$\begin{bmatrix} A & B \\ -B & -A \end{bmatrix} \begin{bmatrix} u \\ v \end{bmatrix} = \lambda \begin{bmatrix} u \\ v \end{bmatrix} \tag{4.2}$$

via a simple orthogonal similarity transformation [2], where A and B are $n \times n$ real symmetric matrices such that the symmetric matrix $\begin{bmatrix} A & B \\ B & A \end{bmatrix}$ is symmetric positive definite[5] [41, 51]. In computational physics and chemistry literature, it is this eigenvalue problem that is referred to as the linear response eigenvalue problem (see, e.g., [36]), or *random phase approximation* (RPA) eigenvalue problem (see, e.g., [15]).

While (4.1) is not a symmetric eigenvalue problem, it has the symmetric structure in its submatrices and many optimization principles that are similar to those one usually finds in the symmetric eigenvalue problem. For example, (4.1) has only real eigenvalues. But more can be said: its eigenvalues come in $\pm\lambda$ pairs. Denote its eigenvalues by

$$-\lambda_n \leq \cdots \leq -\lambda_1 \leq +\lambda_1 \leq \cdots \leq +\lambda_n.$$

[5]This condition is equivalent to that both $A \pm B$ are positive definite. In [2, 3] and this chapter, we focus very much on this case, except that one of $A \pm B$ is allowed to be positive semidefinite.

In practice, the first few positive eigenvalues and their corresponding eigenvectors are needed. In 1961, Thouless [50] obtained a minimization principle for λ_1, now known as *Thouless' minimization principle*, which equivalently stated for (4.1) is

$$\lambda_1 = \min_{x,y} \frac{x^T K x + y^T M y}{2|x^T y|}, \tag{4.3}$$

provided both $K \succ 0$ and $M \succ 0$. This very minimization principle, reminiscent of the first equation in (2.8), has been seen in action recently in, e.g., [8, 31, 33], for calculating λ_1 and, with aid of deflation, other λ_j.

Recently, Bai and Li [2] obtained Ky Fan trace min type principle, as well as Cauchy interlacing inequalities.

Theorem 4.1 (Bai and Li [2]). *Suppose that one of K, $M \in \mathbb{R}^{n \times n}$ is definite.*

1. *We have*

$$\sum_{i=1}^{k} \lambda_i = \frac{1}{2} \inf_{U^T V = I_k} \mathrm{trace}(U^T K U + V^T M V). \tag{4.4}$$

   Moreover, "inf" can be replaced by "min" if and only if both K and M are definite. When they are definite and if also $\lambda_k < \lambda_{k+1}$, then for any U and V that attain the minimum can be used to recovered λ_j for $1 \le j \le k$ and their corresponding eigenvectors.

2. *Let $U, V \in \mathbb{R}^{n \times k}$ such that $U^T V$ is nonsingular. Write $W = U^T V = W_1^T W_2$, where $W_i \in \mathbb{R}^{k \times k}$ are nonsingular, and define*

$$H_{\mathrm{SR}} = \begin{bmatrix} 0 & W_1^{-T} U^T K U W_1^{-1} \\ W_2^{-T} V^T M V W_2^{-1} & 0 \end{bmatrix}. \tag{4.5}$$

   Denote by $\pm \mu_i$ $(1 \le i \le k)$ the eigenvalues of H_{SR}, where $0 \le \mu_1 \le \cdots \le \mu_k$. Then

$$\lambda_i \le \mu_i \le \frac{\sqrt{\min\{\kappa(K), \kappa(M)\}}}{\cos \angle(\mathcal{U}, \mathcal{V})} \lambda_{i+n-k} \quad \text{for } 1 \le i \le k, \tag{4.6}$$

   where $\mathcal{U} = \mathcal{R}(U)$ and $\mathcal{V} = \mathcal{R}(V)$, and $\kappa(K) = \|K\|_2 \|K^{-1}\|_2$ and $\kappa(M) = \|M\|_2 \|M^{-1}\|_2$ are the spectral condition numbers.

Armed with these minimization principles, we can work out extensions of the previously discussed steepest descent methods in subsection 2.3 and conjugate gradient methods in subsection 2.4 for the linear response eigenvalue problem (4.1). In fact, some extensions have been given in [2, 3, 42].

5 Hyperbolic quadratic eigenvalue problem

It was argued in [20] that the hyperbolic quadratic eigenvalue problem (HQEP) is the closest analogue of the standard Hermitian eigenvalue problem $Hx = \lambda x$ when it comes to the quadratic eigenvalue problem

$$(\lambda^2 A + \lambda B + C)x = 0. \qquad (5.1)$$

In many ways, both problems share common properties: the eigenvalues are all real, and for HQEP there is a version of the min-max principles [10] that is very much like the Courant-Fischer min-max principles.

When (5.1) is satisfied for a scalar λ and nonzero vector x, we call λ a *quadratic eigenvalue*, x an associated *quadratic eigenvector*, and (λ, x) a *quadratic eigenpair*.

One source of HQEP (5.1) is dynamical systems with friction, where A, C are associated with the kinetic-energy and potential-energy quadratic form, respectively, and B is associated with the Rayleigh dissipation function. When A, B, and C are Hermitian, and A and B are positive definite and C positive semidefinite, we say the dynamical system is *overdamped* if

$$(x^H Bx)^2 - 4(x^H Ax)(x^H Cx) > 0 \quad \text{for any nonzero vector } x. \qquad (5.2)$$

An HQEP is slightly more general than an overdamped QEP in that B and C are no longer required positive definite or positive semidefinite, respectively. However, a suitable shift in λ can turn an HQEP into an overdamped HQEP [16].

In what follows, A, B, $C \in \mathbb{C}^{n \times n}$ are Hermitian, $A \succ 0$, and (5.2) holds. Thus (5.1) is a HQEP for $\boldsymbol{Q}(\lambda) = \lambda^2 A + \lambda B + C \in \mathbb{C}^{n \times n}$. Denote its quadratic eigenvalues by $\lambda_i^\pm$ and arrange them in the order of

$$\lambda_1^- \leq \cdots \leq \lambda_n^- < \lambda_1^+ \leq \cdots \leq \lambda_n^+. \qquad (5.3)$$

Consider the following equation in λ

$$f(\lambda, x) := x^H \boldsymbol{Q}(\lambda)x = \lambda^2 (x^H Ax) + \lambda(x^H Bx) + (x^H Cx) = 0, \qquad (5.4)$$

given $x \neq 0$. Since $\boldsymbol{Q}(\lambda)$ is hyperbolic, this equation always has two distinct real roots (as functions of x)

$$\rho_\pm(x) = \frac{-x^H Bx \pm \left[(x^H Bx)^2 - 4(x^H Ax)(x^H Cx) \right]^{1/2}}{2(x^H Ax)}. \qquad (5.5)$$

We shall call $\rho_+(x)$ the *pos-type Rayleigh quotient* of $\boldsymbol{Q}(\lambda)$ on x, and $\rho_-(x)$ the *neg-type Rayleigh quotient* of $\boldsymbol{Q}(\lambda)$ on x.

Theorem 5.1 below is a restatement of [32, Theorems 32.10, 32.11 and Remark 32.13]. However, it is essentially due to Duffin [10] whose proof, although for overdamped Q, works for the general hyperbolic case. They can be considered as a generalization of the Courant-Fischer min-max principles (see [38, p.206], [47, p.201]).

Theorem 5.1 ([10]). *We have*

$$\lambda_i^+ = \max_{\substack{\mathcal{X} \subseteq \mathbb{C}^n \\ \text{codim } \mathcal{X}=i-1}} \min_{\substack{x \in \mathcal{X} \\ x \neq 0}} \rho_+(x), \quad \lambda_i^+ = \min_{\substack{\mathcal{X} \subseteq \mathbb{C}^n \\ \dim \mathcal{X}=i}} \max_{\substack{x \in \mathcal{X} \\ x \neq 0}} \rho_+(x), \qquad (5.6a)$$

$$\lambda_i^- = \max_{\substack{\mathcal{X} \subseteq \mathbb{C}^n \\ \text{codim } \mathcal{X}=i-1}} \min_{\substack{x \in \mathcal{X} \\ x \neq 0}} \rho_-(x), \quad \lambda_i^- = \min_{\substack{\mathcal{X} \subseteq \mathbb{C}^n \\ \dim \mathcal{X}=i}} \max_{\substack{x \in \mathcal{X} \\ x \neq 0}} \rho_-(x). \qquad (5.6b)$$

In particular,

$$\lambda_1^+ = \min_{x \neq 0} \rho_+(x), \quad \lambda_n^+ = \max_{x \neq 0} \rho_+(x), \qquad (5.7a)$$

$$\lambda_1^- = \min_{x \neq 0} \rho_-(x), \quad \lambda_n^- = \max_{x \neq 0} \rho_-(x). \qquad (5.7b)$$

To generalize Ky Fan trace min/max principle and Cauchy's interlacing inequalities, we introduce the following notations. For $X \in \mathbb{C}^{n \times k}$ with rank$(X) = k$, $X^H Q(\lambda) X$ is a $k \times k$ hyperbolic quadratic matrix polynomial. Hence its quadratic eigenvalues are real. Denote them by $\lambda_{i,X}^{\pm}$ arranged as

$$\lambda_{1,X}^- \leq \cdots \leq \lambda_{k,X}^- \leq \lambda_{1,X}^+ \leq \cdots \leq \lambda_{k,X}^+.$$

Theorem 5.2.   *1. [28] We have*

$$\min_{\text{rank}(X)=k} \sum_{j=1}^{k} \lambda_{j,X}^{\pm} = \sum_{j=1}^{k} \lambda_j^{\pm}, \quad \max_{\text{rank}(X)=k} \sum_{j=1}^{k} \lambda_{j,X}^{\pm} = \sum_{j=1}^{k} \lambda_{n-k+j}^{\pm}.$$
$$(5.8)$$

2. [52] For $X \in \mathbb{C}^{n \times k}$ with rank$(X) = k$,

$$\lambda_i^+ \leq \lambda_{i,X}^+ \leq \lambda_{i+n-k}^+, \quad i = 1, \cdots, k, \qquad (5.9a)$$

$$\lambda_j^- \leq \lambda_{j,X}^- \leq \lambda_{j+n-k}^-, \quad j = 1, \cdots, k. \qquad (5.9b)$$

Armed with these minimization principles, we can work out extensions of the previously discussed steepest descent methods in subsection 2.3 and conjugate gradient methods in subsection 2.4 for the HQEP $Q(\lambda)x = 0$. Details, among others, can be found in [28].

References

[1] E. Anderson, Zhaojun Bai, C. H. Bischof, S. Blackford, J. W. Demmel, J. J. Dongarra, J. J. Du Croz, A. Greenbaum, S. J. Hammarling, A. McKenney, and D. C. Sorensen. *LAPACK Users' Guide*. Society for Industrial and Applied Mathematics, Philadelphia, PA, USA, 3rd edition, 1999.

[2] Zhaojun Bai and Ren-Cang Li. Minimization principle for linear response eigenvalue problem, I: Theory. *SIAM J. Matrix Anal. Appl.*, 33(4): 1075–1100, 2012.

[3] Zhaojun Bai and Ren-Cang Li. Minimization principle for linear response eigenvalue problem, II: Computation. *SIAM J. Matrix Anal. Appl.*, 34(2): 392–416, 2013.

[4] R. Bhatia. *Matrix Analysis*. Graduate Texts in Mathematics, vol. 169. Springer, New York, 1996.

[5] R. Bhatia. *Positive Definite Matrices*. Princeton Series in Applied Mathematics. Princeton University Press, Princeton, New Jersey, 2007.

[6] P. Binding, B. Najman, and Qiang Ye. A variational principle for eigenvalues of pencils of Hermitian matrices. *Integr. Eq. Oper. Theory*, 35: 398–422, 1999.

[7] W. W. Bradbury and R. Fletcher. New iterative methods for solution of the eigenproblem. *Numer. Math.*, 9(3): 259–267, 1966.

[8] M. Challacombe. Linear scaling solution of the time-dependent self-consisten-field equations. e-print arXiv: 1001.2586v2, 2010.

[9] J. Demmel. *Applied Numerical Linear Algebra*. SIAM, Philadelphia, PA, 1997.

[10] R. Duffin. A minimax theory for overdamped networks. *Indiana Univ. Math. J.*, 4: 221–233, 1955.

[11] R. Fletcher and C. M.Reeves. Function minimization by conjugate gradients. *Comput. J.*, 7: 149–154, 1964.

[12] G. Golub and Qiang Ye. An inverse free preconditioned Krylov subspace methods for symmetric eigenvalue problems. *SIAM J. Sci. Comput.*, 24: 312–334, 2002.

[13] G. H. Golub and C. F. Van Loan. *Matrix Computations*. Johns Hopkins University Press, Baltimore, Maryland, 3rd edition, 1996.

[14] A. Greenbaum. *Iterative Methods for Solving Linear Systems*. SIAM, Philadelphia, 1997.

[15] M. Grüning, A. Marini, and X. Gonze. Implementation and testing of Lanczos-based algorithms for random-phase approximation eigenproblems. Technical report, arXiv:1102.3909v1, February 2011.

[16] C.-H. Guo and P. Lancaster. Algorithms for hyperbolic quadratic eigenvalue problems. *Math. Comp.*, 74: 1777–1791, 2005.

[17] M. R. Hestenes and E. Stiefel. Methods of conjugate gradients for solving linear systems. *J. Res. Nat. Bur. Standards*, 49: 409–436, 1952.

[18] U. Hetmaniuk and R. Lehoucq. Basis selection in LOBPCG. *J. Comput. Phys.*, 218(1): 324–332, 2006.

[19] N. J. Higham. *Functions of Matrices: Theory and Computation*. Society for Industrial and Applied Mathematics, Philadelphia, PA, USA, 2008.

[20] N. J. Higham, F. Tisseur, and P. M. Van Dooren. Detecting a definite Hermitian pair and a hyperbolic or elliptic quadratic eigenvalue problem, and associated nearness problems. *Linear Algebra Appl.*, 351–352: 455–474, 2002.

[21] L. V. Kantorovich and G. P. Akilov. *Functional Analysis in Normed Spaces*. MacMillian, New York, 1964.

[22] A. V. Knyazev and A. L. Skorokhodov. On exact estimates of the convergence rate of the steepest ascent method in the symmetric eigenvalue problem. *Linear Algebra Appl.*, 154-156: 245–257, 1991.

[23] A. V. Knyazev. Toward the optimal preconditioned eigensolver: Locally optimal block preconditioned conjugate gradient method. *SIAM J. Sci. Comput.*, 23(2): 517–541, 2001.

[24] A. V. Knyazev and K. Neymeyr. A geometric theory for preconditioned inverse iteration III: A short and sharp convergence estimate for generalized eigenvalue problems. *Linear Algebra Appl.*, 358(1-3): 95–114, 2003.

[25] J. Kovač-Striko and K. Veselić. Trace minimization and definiteness of symmetric pencils. *Linear Algebra Appl.*, 216: 139–158, 1995.

[26] P. Lancaster and Q. Ye. Variational properties and Rayleigh quotient algorithms for symmetric matrix pencils. *Oper. Theory: Adv. Appl.*, 40: 247–278, 1989.

[27] Xin Liang and Ren-Cang Li. Extensions of Wielandt's min-max principles for positive semi-definite pencils. *Linear and Multilinear Algebra*, 62(8): 1032–1048, 2014.

[28] Xin Liang and Ren-Cang Li. The hyperbolic quadratic eigenvalue problem. Technical Report 2014-01, Department of Mathematics, University of Texas at Arlington, January 2014. Available at http://www.uta.edu/math/preprint/.

[29] Xin Liang, Ren-Cang Li, and Zhaojun Bai. Trace minimization principles for positive semi-definite pencils. *Linear Algebra Appl.*, 438: 3085–3106, 2013.

[30] D. E. Longsine and S. F. McCormick. Simultaneous Rayleigh-quotient minimization methods for $Ax = \lambda Bx$. *Linear Algebra Appl.*, 34: 195–234, 1980.

[31] M. J. Lucero, A. M. N. Niklasson, S. Tretiak, and M. Challacombe. Molecular-orbital-free algorithm for excited states in time-dependent perturbation theory. *J. Chem. Phys.*, 129(6): 064114, 2008.

[32] A. S. Markus. *Introduction to the Spectral Theory of Polynomial Operator Pencils*. Translations of mathematical monographs, vol. 71. AMS, Providence, RI, 1988.

[33] A. Muta, J.-I. Iwata, Y. Hashimoto, and K. Yabana. Solving the RPA eigenvalue equation in real-space. *Progress Theoretical Physics*, 108(6): 1065–1076, 2002.

[34] B. Najman and Qiang Ye. A minimax characterization of eigenvalues of Hermitian pencils II. *Linear Algebra Appl.*, 191: 183–197, 1993.

[35] J. Nocedal and S. Wright. *Numerical Optimization*. Springer, 2nd edition, 2006.

[36] J. Olsen, H. J. Aa. Jensen, and P. Jørgensen. Solution of the large matrix equations which occur in response theory. *J. Comput. Phys.*, 74(2): 265–282, 1988.

[37] E. E. Ovtchinnikov. Sharp convergence estimates for the preconditioned steepest descent method for Hermitian eigenvalue problems. *SIAM J. Numer. Anal.*, 43(6): 2668–2689, 2006.

[38] B. N. Parlett. *The Symmetric Eigenvalue Problem*. SIAM, Philadelphia, 1998.

[39] B. T. Polyak. *Introduction to Optimization*. Optimization Software, New York, 1987.

[40] P. Quillen and Qiang Ye. A block inverse-free preconditioned Krylov subspace method for symmetric generalized eigenvalue problems. *J. Comput. Appl. Math.*, 233(5): 1298–1313, 2010.

[41] P. Ring and P. Schuck. *The Nuclear Many-Body Problem*. Springer-Verlag, New York, 1980.

[42] D. Rocca, Z. Bai, R.-C. Li, and G. Galli. A block variational procedure for the iterative diagonalization of non-Hermitian random-phase approximation matrices. *J. Chem. Phys.*, 136: 034111, 2012.

[43] Y. Saad. On the rates of convergence of the Lanczos and the block-Lanczos methods. *SIAM J. Numer. Anal.*, 15(5): 687–706, October 1980.

[44] Y. Saad. *Iterative Methods for Sparse Linear Systems*. SIAM, Philadelphia, 2nd edition, 2003.

[45] B. Samokish. The steepest descent method for an eigenvalue problem with semi-bounded operators. *Izv. Vyssh. Uchebn. Zaved. Mat.*, 5: 105–114, 1958. in Russian.

[46] G. W. Stewart. *Matrix Algorithms, Vol. II: Eigensystems*. SIAM, Philadelphia, 2001.

[47] G. W. Stewart and Ji-Guang Sun. *Matrix Perturbation Theory*. Academic Press, Boston, 1990.

[48] Wenyu Sun and Ya-Xiang Yuan. *Optimization Theory and Methods – Nonlinear Programming*. Springer, New York, 2006.

[49] I. Takahashi. A note on the conjugate gradient method. *Inform. Process. Japan*, 5: 45–49, 1965.

[50] D. J. Thouless. Vibrational states of nuclei in the random phase approximation. *Nuclear Physics*, 22(1): 78–95, 1961.

[51] D. J. Thouless. *The Quantum Mechanics of Many-Body Systems*. Academic, 1972.

[52] K. Veselić. Note on interlacing for hyperbolic quadratic pencils. In Jussi Behrndt, Karl-Heinz Förster, and Carsten Trunk, editors, *Recent Advances in Operator Theory in Hilbert and Krein Spaces*, volume 198 of *Oper. Theory: Adv. Appl.*, pages 305–307. 2010.

[53] Qiang Ye. *Variational Principles and Numerical Algorithms for Symmetric Matrix Pencils*. PhD thesis, University of Calgary, Calgary, Canada, 1989.

Factorization-Based Sparse Solvers and Preconditioners

Xiaoye Sherry Li *

Abstract

Efficient solution of large-scale, ill-conditioned and highly-indefinite algebraic equations often relies on high quality preconditioners together with iterative solvers. Because of their robustness, factorization-based algorithms play a significant role in developing scalable solvers. We discuss the state-of-the-art, high performance sparse factorization techniques which are used to build sparse direct solvers, domain-decomposition type direct/iterative hybrid solvers, and approximate factorization preconditioners. In addition to algorithmic principles, we also address the key parallelism issues and practical aspects that need to be taken under consideration in order to deliver high speed and robustness to the users of today's sophisticated high performance computers.

1 Fundamentals of parallel computing

Parallel computing has become an increasingly indispensable tool in various computing disciplines, such as modeling physical phenomena in science and engineering simulations as well as technical computing in industry. Here, we give a brief overview of the parallel architectures, programming, applications, and performance. The more thorough treatment of the subject can be found in [7, 62] and many references therein.

1.1 Parallel architectures and programming

Parallelism is ubiquitous and occurs at many levels of hierarchy in modern processor architectures. There are several different forms of parallel computing with varying granualities: bit level, instruction level, data, and task parallelism. *Pipelining* is the most fundamental form of parallelism commonly used to increase throughput. In a pipeline, the computation for an input is divided into stages with each stage running on its own spatial division of the processors. The output of one stage is the

---

*Lawrence Berkeley National Laboratory, USA, xsli@lbl.gov.

input of the next one. The different stages of the pipeline are often executed in parallel or in time-sliced fashion. The basic usages of pipeline are instruction execution, arithmetic computation and memory access. For example, the instruction circuitry can be divided into five stages: instruction fetch, instruction decode, register fetch, arithmetic, and register write back stages, wherein each stage processes one instruction at a time. This allows overlapping execution of five instructions at different stages.

Based on the pipeline principle, the *vector machines* with SIMD (Single Instruction, Multiple Data) instructions were introduced to provide high-level operations on arrays of elements. SIMD instructions perform exactly the same operations on multiple data objects, thus produce multiple results at the same time. Each instruction pipelines the operations on the individual elements of a vector. The pipeline includes the arithmetic operations (e.g., multiplication, addition, etc.) and memory access. For example, in 1999, Intel introduced the SSE (Streaming SIMD Extensions) (and SSE2 extension for double precision) in the x86 architectures. The hardware is augmented with a set of vector registers, each of length 128 bits. Each floating-point vector instruction can deliver four results of four pairs of single-precision numbers or two results of two pairs of double-precision numbers. Intel's newer SIMD instruction set is called AVX which supports 512-bit wide vectors [9]. The other examples of SIMD instruction sets include VIS (Sun Microsystem's SPARC) and AltiVec (Apple/IBM/Freescale Semiconductor).

In recent years, GPU computing has become commonplace in scientific and enginnering applications. Here, a GPU (graphics processing unit) is used together with a CPU to accelerate computations. CPU + GPU is a powerful combination because CPUs consist of a few cores optimized for serial processing, while GPUs consist of thousands of smaller, more efficient cores designed for applications with abundance of *data parallelism*. This provides unprecedented application performance by offloading compute-intensive portions of the application to the GPU, while the remainder of the code still runs on the CPU.

The state-of-the-art massively parallel architectures usually consist of clusters of distributed memory, manycore nodes. For example, in the list of the world's top 500 fastest supercomputers (http://www.top500.org/), the first on the list is **Tianhe-2**, which contains 16,000 compute nodes, each comprising two Intel Ivy Bridge Xeon processors and three Xeon Phi chips (with wide SIMD paralleism). The second on the list is **Titan**, which contains 18,688 compute nodes, each comprising a 16-core AMD processor and an Nvidia Kelper GPU. Utilizing such machines requires exploiting both the coarse-level task parallelism and fine-grained data parallelism, wherein the dataset is divided into pieces each of which is stored on one compute-node. Data-parallel computations can be performed locally on each node using the locally stored data. When needed, nodes may send data to the other nodes through the interconnect fabric for cooperatively performing the same task.

The level of sophistication of programming the parallel machines above varies with different forms of parallelism. For the small degree parallelism provided by the pipeline and vector forms, the compilers can usually generate efficient codes to make full use of the hardware features. The users do not need to write explicit parallel programs. For the moderate parallelism provided by the shared memory machines, a commonly used standard programming is OpenMP [85, 86]. OpenMP is an implementation of multithreading, whereby a master thread (a series of instructions executed consecutively) forks a specified number of slave threads and a task is divided among them. The threads then run concurrently, with the runtime system allocating threads to different processors.

For the massive parallelism provided by the distributed memory machines, the commonly used standard programming is MPI (Message Passing Interface) [80, 81]. MPI primarily addresses the message-passing parallel programming model: data is partitioned and distributed among the address spaces of different processes. Through cooperative operations, data is transferred from the address space of one process to that of another process.

For the cluster of distributed memory with heterogeneous nodes with multicores and/or GPUs, it is insufficient to use MPI alone. We need to use a hybrid programming model such as MPI+X, where MPI is used to generate multiple processes across multiple nodes, and X is used within the address space of each MPI process. Here, X can be OpenMP for the classic cache-based multicore processors or OpenCL [84] (or Nvidia's CUDA [25]) for the GPUs on the node.

1.2   Parallel algorithms and applications

Vast amounts of applications can benefit from different levels of parallelisms. In the early days of parallel computing, different parallel algorithms were designed and tailored for different applications. This limits the reusability of the algorithms and codes. A sustainable approach is to identify a number of patterns of communication and computation each of which is *common* to a class of applications. Colella first proposed a high level of abstraction categorizing seven such patterns (*Seven Dwarves*) for the numerical methods commonly used in scientific and engineering applications [24]: Structured grids, Unstructured grids, Dense linear algebra, Sparse linear algebra, FFT, N-body methods, and Monte Carlo. Later, the researchers at Berkeley extended the list to *Thirteen Dwarves* to capture the parallel computing patterns in

broader applications [7]. The additional six dwarves are: Combinatorial
logic, Graph traversal, Dynamic programming, Backtrack and branch-
and-bound, Construct graphical models, and Finite state machine. The
parallel hardware, software and algorithms can be designed to optimize
performance for each dwarf.

1.3   Performance models and upper bounds

Given the complexity of the modern architectures, the actual runtime
of the parallel algorithms and codes are extremely difficult to estimate.
Despite this, the performance upper bounds can be predicted reasonably
well using the following perofrmance models: the roofline model, Amdahl's law, and the latency-bandwidth model. We briefly present them
below.

In numerical computing the traditional metric of analyzing an algorithm efficiency is flop count. This is far from an accurate performance
predictor for the modern high-performance machines. Now performance
is more dominated by memory access and inter-node communication,
particularly for the algorithms involving graphs and sparse matrices.
Arithmetic Intensity (AI) is a measure to capture both floating-point
operations and the memory/network traffic; it is calculated as the ratio
of floating-point operations to DRAM traffic in bytes, i.e., flops:bytes
ratio. For example, the AI for Level 3 BLAS is $O(n)$ and for Levels
1 and 2 BLAS is $O(1)$, where n is the matrix dimension. The AI for
FFT is $O(\log n)$, where n is the number of points. A higher AI indicates
more potential for data reuse in cache, and the kernel is more amenable
to various code optimizations to achieve a higher percent of machine's
peak performance. The *roofline model* gives a more realistic performance upper bound depending on both the computer architecture and
the algorithm/code to be executed [109]. Simply put, the performance
is bound to

$$\text{Attainable Performance} = \min \begin{cases} \text{Peak FLOP performance} \\ \text{Peak Bandwidth} \times AI \end{cases}.$$

Figure 1.1 depicts the peak roofline ceiling as a function of AI. Depending on how extensive the various optimization techniques are used,
the actual roofline may be lower than the peak. For example, if the
SSE2 instruction (128-bit SIMD) is not used on the Intel Opterons, the
peak FLOP ceiling would be halved. Similarly for memory performance,
if the code exhibits many random memory access patterns, the peak
bandwidth ceiling would be lower.

For parallel computations *speedup* is commonly used to measure the
performance gains achieved by using multiple processors, which is defined
as the ratio of the sequential runtime over the parallel runtime. *Amdahl's*

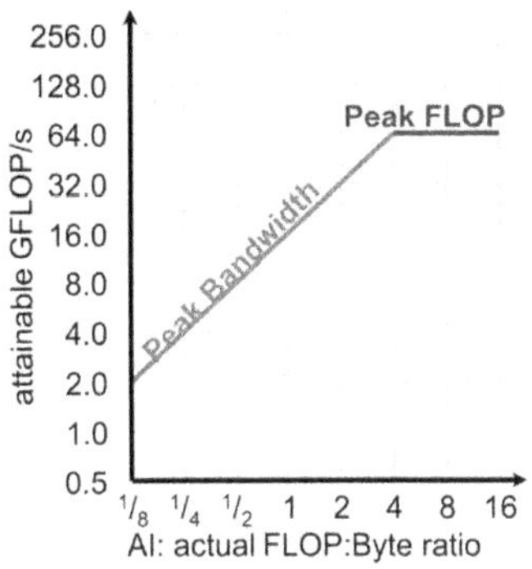

Figure 1.1   The roofline performance bound of a hyperthetocal machine (source: S. Williams)

law gives an upper bound of the attainable speedup of a given parallel algorithm [1], independent of the machine architecture. In a parallel application, let s be the fraction of the work performed sequentially, $1 - s$ is the fraction parallelized, and P is the number of cores. Then,

$$\text{Attainable Speedup} = \frac{1}{s + \frac{1-s}{P}} \leq \frac{1}{s}.$$

That is, the sequential chunk of work prevents the code from scaling up no matter how many cores are used. Therefore, nearly 100% of the code needs to be parallelized in order to use millions of cores.

On the distributed memory systems, it is necessary to model the cost of transferring data between different nodes, e.g., using MPI. A commonly used cost model for network performance is α-β *model*, where α refers to the latency and β is the inverse of the bandwidth between two processors. The time to send a message of length n is roughly:

$$\text{Time} = \text{latency} + n/\text{bandwidth} = \alpha + n \times \beta .$$

This is a simplified and ideal model, without taking into account such practicalities as network congestion etc. For most parallel machines we have $\alpha \gg \beta \gg \text{time\_per\_flop}$. For example, on a Cray XE6, $\text{time\_per\_flop} = 0.11\text{ns}$. Using MPI message transfer, $\alpha = 1.5\mu\text{s} \approx 13,636$ flops, and $\beta = 0.17\text{ns} \approx 12$ flops_per_double_word. Therefore, the fundamental principle is to organize the parallel algorithm so that it maintains high data locality and sends fewer long messages rather than sending many short messages.

2   Sparse matrix basics

Sparse matrices are ubiquitous in scientific and engineering calculations. A matrix is considered sparse if there are many zeros in it, and it is worth using special algorithms to perform matrix operations on it. A large class of sparse matrices arise from discretizing partial differential equations (PDE) for which the number of nonzeros in the matrix does not grow proportionally to the square of the matrix dimension (n^2), but only grows linearly w.r.t. n. Therefore, when the problem size increases, the sparsity #nonzeros$/n^2$ becomes smaller and sparse matrix algorithms have asymptotically lower complexity than the dense counterparts.

The first issue to address is the data structure used to store a sparse matrix. The goal is to store only the nonzeros in the matrix and to perform operations only on the nonzeros, and to handle arbitrary sparsity patterns. In the early days, the *coordinate* format (a.k.a. triplets) and the *linked list* were used. Although flexible, they are not efficient for many matrix algorithms on high performance computers — the former requires more storage than necessary and the latter prevents the algorithm from using the BLAS routines directly. Today several other compressed formats are more widely used. The most popular format is *Compressed Row Storage (CRS)* or *Compressed Column Storage (CCS)*. Let n denote the dimension of the matrix and nnz denote the number of nonzeros in the matrix. The CSR format consists of three vectors: one for floating-point numbers and the other two for integer values. The nzval vector of size nnz stores the nonzero values row-by-row contiguously. The colind vector of size nnz stores the column indices of the entries in nzval. The rowptr vector of size $n + 1$ stores the position in nzval that starts a new row. For example, the following 7-by-7 sparse matrix

$$\begin{bmatrix} 1 & & & a & & & \\ & 2 & & b & & & \\ c & d & 3 & & & & \\ & e & & 4 & f & & \\ & & & & 5 & & g \\ & & & h & i & 6 & j \\ & & k & & l & & 7 \end{bmatrix}$$

is represented in CRS as follows:

nzval	1	a	2	b	c	d	3	e	4	f	5	g	h	i	6	j	k	l	7
colind	1	4	2	5	1	2	3	2	4	5	5	7	4	5	6	7	3	5	7

rowptr	1	3	5	8	11	13	17	20

.

The CCS (a.k.a. Harwell-Boeing) format is a symmetric analogue of CRS as follows:

nzval	1	c	2	d	e	3	k	a	4	h	b	f	5	i	l	6	g	j	7
rowind	1	3	2	3	4	3	7	1	4	6	2	4	5	6	7	6	5	6	7

colptr	1	3	6	8	11	16	17	20

Both formats require storage of nnz floating-point numbers and $nnz + n + 1$ integers. This is smaller than that required by the coordinate format: nnz floating-point numbers and $2 \times nnz$ integers. With the CRS storage format, the algorithm needs to traverse the matrix in a row-wise fashion to ensure sequential access to the nzval and colind arrays, whereas with the CCS format, the algorithm need to traverse the matrix in a column-wise fashion.

The other sparse data structures include block-entry format, skyline or profile format, ELLPACK format and segmented-sum format, see [11] for details. The latter two are good for machines with wide SIMD instructions.

In sparse iterative solution methods, the most used operation is sparse matrix-vector multiplication (SpMV): $y = Ax$. Using the aforementioned representations for A, it is fairly straightforward to code the algorithms for $y = Ax$. However, the performance of the simple algorithms is usually rather low. The main bottlenecks are due to the low arithmetic intensity (bandwidth bound) and the strided access to the x or y vectors. A number of optimization techniques have been developed to mitigate the problems which can achieve several fold speedup relative to the baseline implementations, see [12] for many papers on this topic.

A unique aspect of sparse matrix computations is the connection to combinatorial algorithms related to graphs. An indispensible tool is the graph manipulation to reason about the nonzero structure and to transform (e.g. via reordering) the matrix to increase the performance of the numerical computation. A graph $G = (V, E)$ consists of a finite set V, called the vertex set and a finite, binary relation E on V, called the edge set. There are three standard graph models commonly used for sparse matrices.

- *Undirected graph:* The edges are unordered pairs of vertices, that is, $\{u, v\} \in E$ for some $u, v \in V$; Undirected graphs can be used to represent symmetric matrices: rows/columns correspond to the vertex set V. For each nonzero $A(i, j)$ there is an edge $\{v_i, v_j\}$.

- *Directed graph:* The edges are ordered pairs of vertices, that is, (u, v) and (v, u) are two different edges; Directed graphs can be used to represent nonsymmetric matrices.

- *Bipartite graph:* $G = (U \cup V; E)$ consists of two disjoint vertex sets U and V such that for each edge $\{u, v\} \in E, u \in U$ and $v \in V$. Bipartite graphs can be used to represent rectangular or nonsymmetric matrices.

The *degree* of a vertex v is the number of neighboring vertices connected to v. An *ordering* or labeling of $G = (V, E)$ having n vertices, i.e., $|V| = n$, is a mapping of V onto $\{1, 2, \ldots, n\}$. Very often, the sparse matrix directly coming from the physical model is not in the best ordering. We can apply various transformations by reordering the rows (equations) and columns (variables) of the matrix (linear system), which serve different purposes in different operations. In SpMV $y = Ax$, a reordering may improve access locality to the x or y vectors, and reduce communication in a parallel algorithm (see e.g. [83, 102, 104, 110].) In sparse LU factorization $A = LU$, a reordering may reduce the number of fill-ins in the L and U factored matrices (see e.g. [3, 5, 42, 45, 65, 78, 23]).

3 Direct methods for sparse linear systems

Direct methods for solving a sparse linear system $Ax = b$ are based on Gaussian elimination (GE). The matrix A is first decomposed (factorized) into a lower triangular matrix L and an upper triangular matrix U, then x is obtained by forward substitution with L followed by back substitution with U. When A is symmetric and positive definite (SPD), the Cholesky factorization $A = LL^T$ can be computed. When A is symmetric and indefinite, LDL^T can be computed. In both cases, saving is obtained by exploiting symmetry. In many situations, we need to solve a transformed linear system for accuracy and/or performance reasons. When A is dense we often use *partial pivoting* during GE and the resulting factorization is $PA = LU$, where P is a permutation matrix determined such that at each step of elimination the largest-magnitude entry of the column is chosen as the pivot and the corresponding row is swapped to the pivot row. Sometimes *complete pivoting* is used to swap the largest-magnitude entry of the entire trailing submatrix to the pivot position, resulting in a factorization $PAQ = LU$ with both rows and columns being permuted.

Many more complicated issues arise in sparse factorizations because of the *fill-ins*, which are the new nonzeros generated in the factored matrices L and U. Figure 3.1 shows a 7×7 matrix with the original nonzeros in black dots and the fill-ins in red dots.

A typical sparse solver consists of the following four distinct steps:

1. An *ordering* step that reorders the rows and columns such that the factors suffer little fill, or that the matrix has special structure such as block triangular form [37, 91].

2. An analysis step or *symbolic factorization* that determines the nonzero structures of the factors and create suitable data structures for the factors.

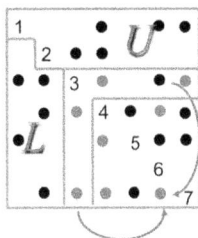

Figure 3.1   Illustration of the fill-ins in sparse GE.

3. A numerical factorization step that computes the L and U factors.

4. A triangular solution step that performs forward and back substitution using the factors.

There are vast varieties of algorithms associated with each step. Usually steps 1 and 2 involve only the graphs of the matrices and integer operations. Steps 3 and 4 involve floating-point operations. Step 3 is often the most time-consuming part, whereas step 4 can be orders of magnitude faster. The algorithm used in step 1 is quite independent of that used in step 3. But the algorithm in step 2 is often closely related to that of step 3. In a solver for SPD systems, the four steps can be well separated. For the most general unsymmetric systems, the solver may combine steps 2 and 3 (e.g. SuperLU) or even combine steps 1, 2 and 3 (e.g. UMFPACK) so that the numerical values also play a role in determining the elimination order.

3.1   Combinatorics

The ordering algorithms and symbolic factorizations are based on various graph models. They are simpler and faster for symmetric factorizations (e.g. Cholesky factorization) than for unsymmetric factorizations. Given certain elimination order $(v_1, v_2, \ldots, v_n)$, the purpose of the symbolic analysis is to determine the fill-in positions for the factors L and U (called the *filled graph*, denoted as $G^+(A)$), and set up the sparse compressed data structures for them. It turns out that the fill-ins can often be discovered solely based on the original graph $G(A)$. The graph tool to aid this task is *reachable set*. A node v_j is reachable from node v_i if there is a path from v_i to v_j in the graph. Let S be a subset of the vertex set, the reachable set of y through S consists of the set of vertices x such that x is reachable from y via a path $(y, v_1, \ldots, v_k, x)$ and all the intermediate vertices v_is are in S; this is denoted as $Reach(y, S)$. The following fill-path theorem by Rose and Tarjan gives precisely the edges in $G^+(A)$.

Theorem 3.1. *[94] Let $G(A) = (V, E)$ be a directed graph of A, then an edge (v, w) exists in the filled graph $G^+(A)$ if and only if $w \in Reach(v, \{v_1, \ldots, v_k\})$, where $v_i < \min(v, w), 1 \le i \le k$.*

For a Cholesky factorization LL^T, the graph reachability can be computed efficiently with the aid of *elimination tree* (or etree) [99, 74]: The vertices of the tree are integers 1 through n, representing the columns of A. The first nonzero $L(j, i), j > i$ in column i defines a child-parent edge (i, j) of the tree. The etree can be computed based solely on $G(A)$ in almost linear time. The symbolic factorization algorithm can simply traverse the etree bottom-up to discover all the reachable vertices (nonzeros) from the nonzeros in the children's columns. The computational complexity is linear w.r.t. the number of nonzeros in the factor L. For unsymmetric LU factorization, the symbolic factorization involves more work. The unsymmetric analogue of the etree is the *column elimination tree*, i.e. the etree of $A^T A$ (or column etree). However, the analysis based on the column etree only gives the nonzero structure of the Cholesky factor of $A^T A$, which is an upper bound on the nonzero structure of L and U^T in $A = LU$. The precise symbolic factorization for LU has to base on the elimination DAGs of both factors L and U [49]. The computational complexity is more than linear w.r.t. the number of nonzeros in L and U, but is much lower than the flop count needed for LU factorization when employing symmetric pruning [39] and supernodes [29].

The sparsity-preserving ordering is an important research area. The goal is to solve an optimization problem of finding the best ordering of equations and variables so that the number of fill-ins in the factors is minimized. Unfortunately, finding the optimal ordering is an NP-complete problem [117]. Therefore, many heuristic algorithms for finding a good ordering have been developed. One class of algorithms is called *minimum degree* heuristic [45] for symmetric matrices. This is based on the following graph changes due to elimination: eliminating a variable/equation acts like eliminating a vertex in the associated undirected graph, and the neighboring vertices of this vertex become fully connected (called a *clique*). An edge in that clique would be a fill-in if it is not in A. Therefore, the minimium degree algorithm chooses a vertex with the smallest degree to eliminate next, which minimizes the upper bound on the fill-ins produced at that step (*local greedy* strategy). Although the basic principle is simple, the straightforward implementation is slow and requires too much memory. The main innovation for efficient implementation is the introduction of the *quotient graph* [40, 47], which represents the collection of cliques compactly throughout the elimination in space bounded by the size of $G(A)$ instead of $G^+(A)$. Another idea is to use *approximate degree* which is faster to compute than the exact degree [3, 2]. There are large numbers of research papers on the algorithmic and implementational variants and performance comparison [8, 3, 34, 47, 73, 101], In particular, George and Liu's article presented an excellent review on this subject [45].

In contrast to minimum degree, the *nested dissection* algorithm is a *global* approach based on divide and conquer paradigm. It recursively partitions the (sub)mesh via *separators*. At each level of dissection, the equations/variables associated with the two (or more) parts are first eliminated before those of the separator. From reachability argument, the vertices are not reachable between the two parts because they are separated by the higher numbered separator vertices, therefore no fill-in is generated between the two parts. At the end of recursion, the final ordering is performed such that the lower level separator nodes are ordered before the upper level ones. The recursive bisection procedure results in a complete binary (amalgamated) elimination tree, a.k.a. *separator tree*. George first introduced this for two-dimensional finite element mesh [42], and proved its *optimality* for fill-reduction: the number of nonzeros in the Cholesky factor is $O(n \log n)$ and the operation count is $O(n^{3/2})$. This optimality condition cannot be proven for the minimum degree ordering algorithm. The generalization of nested dissection to the irregular geometry is the *graph partitioning* method [72, 48]. A good algorithm is to find the separators as small as possible, so that the zero-block between the two (sub)parts are large which better preserves sparsity. There is a large body of research on finding small separators [60, 61, 65, 23, 92, 10, 18]. Several good graph partitioning packages have been developed and are publically available, including Chaco [61], ParMetis [64, 66] and PT-Scotch [89, 88].

For unsymmetric LU factorization, a local greedy strategy analogous to minimum degree was developed by Markowitz [78]: At each step of elimination, row and column permutations are performed so as to minimize the product of the number of off-diagonal nonzeros in the pivot row and pivot column (called Markowitz count). This directly minimizes the arithmetic operations and tends to minimize the number of fill-ins. Although effective, it is difficult to implement the Markowitz algorithm efficiently primarily due to two obstacles: one is the lack of compact representation like the quotient graph used in the minimum degree algorithm, and the other is the need for numerical pivoting in LU factorization which requires the Markowitz ordering algorithm to be interleaved with the numerical factorization (see MA28 [32] and MA48 [35]).

Several alternative strategies have been developed to make the ordering for LU separate from the numerical phase so that the implementation of the ordering is more efficient. When the pivots can be chosen on the diagonal, the Markowitz scheme can be implemented efficiently by using the *bipartite quotient graph* [87] and the *bi-clique* [5]. In addition, *local symmetrization* was introduced to restrict the search path of length bounded by three while searching the reachable set to update the Markowitz count. Thus the algorithm can be implemented in space bounded by the size of $G(A)$, and has the same time complexity as that of the minimum degree algorithm [5, 6].

When partial pivoting is needed, one efficient approach is to symmetrize the matrix first forming $A^T A$ (ignoring numerial cancellation), then to apply *any* symmetric ordering algorithm to $G(A^T A)$ which gives a fill-reducing permutation Q. Then, Q is applied to the columns of A before performing the LU decomposition with row pivoting: $PAQ^T = LU$. The rationale behind this is due to the following result.

Theorem 3.2 ([46]). *Consider the Cholesky factorization $A^T A = R^T R$ and the LU factorization with partial pivoting $PA = LU$. For any row permutation P, $struct(L) \subset struct(R^T)$ and $struct(U) \subset struct(R)$.*

Therefore, the nonzero structure of Cholesky factor R_q in $(AQ^T)^T \cdot (AQ^T) = R_q^T R_q$ upper bounds that of L_q^T and U_q in $P(AQ^T) = L_q U_q$. Since with a good fill-reducing ordering Q, R_q contains less fill than R does, which leads to an indirect effect that L_q and U_q are likely to contain less fill than L and U. In essence, the column ordering Q tends to minimize an upper bound on the actual fill-ins in the LU factors, taking into account all the possible row permutations.

The sequential and shared-memory SuperLU solvers use this principle for sparsity ordering. There are also efficient ordering algorithms that are based on the graph $G(A^T A)$, but without forming the matrix $A^T A$ and base solely on $G(A)$: COLAMD is a minimum degree variant [28] and HUND is a nested dissection variant using the hypergraph model for unsymmetric matrices [51].

In general, the minimum degree algorithms work well for small to medium sized problems, while the nest dissection variants work better for large-scale problems. Researchers have developed the *hybrid* methods that perform nested dissection for a few levels, then uses a minimum degree algorithm for each subgraph at the bottom level of dissection [90]. Apart from the two broad classes of methods above, several other ordering algorithms were also used to permute the matrix into certain special forms, such as the maximum matching algorithm to compute the block triangular form [37, 91], the (reverse) Cuthill–McKee algorithms (CM and RCM) to reduce the bandwidth of the nonzero pattern [26, 44].

3.2   Dataflow organization, task ordering

The Gaussian elimination algorithm can be organized in different ways, such as left-looking (fan-in) or right-looking (fan-out). These variants are mathematically equivalent under the assumption that the floating-point operations are associative (approximately true), but they have very different memory access and communication patterns. The pseudo-code for the left-looking blocking algorithm is given in Algorithm 1.

Algorithm 1. *Left-looking Gaussian elimination*

> **for** *block* $K = 1$ **to** N **do**
>> (1) *Compute* $U(1 : K - 1, K)$
>> *(via a sequence of triangular solves)*
>> (2) *Update* $A(K : N, K) \leftarrow A(K : N, K) - L(1 : N, 1 : K - 1) \cdot$
>> $U(1 : K - 1, K)$ *(via a sequence of calls to GEMM)*
>> (3) *Factorize* $A(K : N, K) \rightarrow L(K : N, K)$
>> *(may involve pivoting)*
> **end for**

SuperLU and SuperLU_MT use the left-looking algorithm, which has the following advantages:

- In each step, the sparsity changes are restricted within the Kth block column instead of the entire trailing submatrix, which makes it relatively easy to accommodate dynamic compressed data structures due to partial pivoting.

- There are more memory *read* operations than *write* operations in Algorithm 1. This is better for most modern cache-based computer architectures, because write tends to be more expensive in order to maintain cache coherence.

The pseudo-code for the right-looking blocking algorithm is given in Algorithm 2.

Algorithm 2. *Right-looking Gaussian elimination*

> **for** *block* $K = 1$ **to** N **do**
>> (1) *Factorize* $A(K : N, K) \rightarrow L(K : N, K)$
>> *(may involve pivoting)*
>> (2) *Compute* $U(K, K + 1 : N)$
>> *(via a sequence of triangular solves)*
>> (3) *Update* $A(K + 1 : N, K + 1 : N) \leftarrow$
>> $A(K + 1 : N, K + 1 : N) - L(K + 1 : N, K) \cdot U(K, K + 1 : N)$
>> *(via a sequence of calls to GEMM)*
> **end for**

SuperLU_DIST uses right-looking algorithm mainly for scalability consideration.

- The sparsity pattern and data structure can be determined before numerical factorization because of static pivoting.

- The right-looking algorithm fundamentally has more parallelism: at step (3) of Algorithm 2, all the GEMM updates to the trailing submatrix are independent and so can be done in parallel. On the other hand, each step of the left-looking algorithm involves operations that need to be carefully sequenced, which requires a sophisticated pipelining mechanism to exploit parallelism across multiple loop steps.

The multifrontal method is a variant of right-looking approach [36, 75]. Similar to the right-looking algorithm, at each elimination step, several variables (fully-summed variables) in the supernode (frontal matrix) are eliminated and the Schur complement update matrix is produced. In contrast to the right-looking algorithm where the Schur complement is updated *in place* immediately at each step, the multifrontal method postponed the update until it is time to eliminate the variables that are affected by this update matrix. Operationally, there are a number of such update matrices which are temporarily merged among themselves and stored in memory before they are finally updated into the destination Schur complement location. In other words, the Schur complement needs to be updated by a number of update matrices. The right-looking algorithm performs each update right away, whereas the multifrontal algorithm accumulates the update matrices in the *partial sums* first, and performs the updates to destination at a later stage. The multifrontal method has several advantages: it consists of a sequence of dense matrix operations, and can use Level 3 BLAS to great extent; the partial sum of the update matrices is simply passed only between a node and its parent in the elimination/assembly tree, which eases parallel algorithm design. The biggest drawback is the large memory demand for storing the intermediate update matrices. This storage is often referred to as stack memory.

3.3   Parallelization

A great deal of effort has been invested on parallelizing the numerical factorization and triangular solution phases, because they often contribute to over 90% of the total runtime. The (column) elimination tree is a valuable tool to help design the parallel algorithms. The eliminations of the (super)nodes corresponding to the different branches of the tree can proceed independently in parallel. For the two (super)nodes situated along the same leaf-to-root path, there is a potential dependency due to the update from the descendant node to the ancestral node. Synchronization is needed among the cores owning these nodes in order to preserve the precedence relation during the computation. Intuitively, the sparse factorization should exhibit more parallelism than the dense counterpart, because both the tree-based parallelism and the fine-grained dense matrix parallelism are available. However, the parallel scaling of a sparse factorization is often hampered by many factors, including high communication-to-computation ratio and irregular communication pattern implicitly encoded in the sparse LU DAG.

On a shared memory machine there is no need to partition the matrices. Usually the parallel elimination can use an asynchronous and barrier-free algorithm to schedule different types of tasks to achieve a high degree of concurrency, such as panel or frontal matrix factorization, Schur complement update, or assembly (extend-add) of the update matrices. The scheduler facilitates synchronization among different tasks to preserve task dependency and to maintain dynamic load balance. The example codes include MA41 [4], PARDISO [98], SuperLU_MT [30] and SuiteSparseQR [27]. SuperLU_MT achieved over 10-fold speedups on a number of earlier SMP machines with 16 processors [30]. Recent evaluation shows that SuperLU_MT performs very well on current multithreaded, multicore machines; it achieved over 20-fold speedup on a 16 core, 128 thread Sun VictoriaFalls [69].

The design of a distributed memory algorithm can be drastically different from a shared memory one. Many design choices are made due to the need for scalability on a large number of processors. The input sparse matrix A is usually divided by block rows, with each process having one block row represented in a local row-compressed format. This format is user-friendly and is compatible with the input interface of many other distributed memory sparse matrix software. On the other hand, the factored matrices or the intermediate frontal matrices are often distributed by a two-dimensional block cyclic layout, see e.g., SuperLU_DIST [70] and WSMP [55]. This distribution ensures that most (if not all) processors can participate in the update at each block elimination step, and also ensures that inter-process communication is restricted among the row sets or the column sets of the processes. The right-looking sparse LU factorization in SuperLU_DIST uses elimination DAGs to identify task and data dependencies, and a pipelined look-ahead scheme to overlap communication with computation. SuperLU_DIST has achieved 50- to 100-fold speedups with sufficiently large matrices, and over half a teraflops factorization rate [114].

Although the ordering and the symbolic factorization algorithms require much less time than the numerical algorithms, it is still essential to develop the distributed memory algorithms for memory scalability, because the problem size is becoming too large to fit in the memory of a single node. Parallelizing the minimum degree type of algorithm is very challenging, even for shared memory machines [22]. The nested dissection variants of algorithms exhibit more parallelism, and the notable successes with distributed memory parallel implementations include ParMETIS [66], PT-Scotch [23] and Zoltan [119].

For sparse Cholesky factorization LL^T with SPD matrices, the parallel symbolic factorization algorithm is usually designed based on the elimination tree [54, 43, 68]. The unsymmetric factorization has to base on a more complex graph model—the elimination DAGs [49]. A good parallel algorithm is presented in [52].

3.4　Available software

Researchers had developed a number of sparse direct solver packages throughout many years, which span the spectrum of different factorization methods (e.g., LU, Cholesky, LDLT, QR), and on different parallel machines (e.g., shared memory, distributed memory). The following survey article contains a table of representative codes: `http://crd-legacy.lbl.gov/~xiaoye/SuperLU/SparseDirectSurvey.pdf`. Recently, there have been several research papers on sparse factorizations exploiting the GPU power [67, 76, 118, 97]. We will update our table as new codes become available.

4　Approximate factorizations as preconditioners

When iterative methods are used to solve the linear system, it is often necessary to solve a transformed linear system $M^{-1}Ax = M^{-1}b$, where M is called the preconditioner. A good preconditioner M would approximate A very well so that the eigenvalue distribution of $M^{-1}A$ is better than that of A. On the other hand, we would like M to be cheap to compute and to invert. Very often the two objectives are contradicting and trade-off is desired in practice. There is a large body of research on preconditioners, see the excellent survey by Benzi [14]. Here, we present two classes of approximate factorization methods that are based on "exact" factorization, but remove some "small" entries in the factors and hence achieve better time and memory efficiency. The approximate factorizations can be used as "black-box" algebraic preconditioners for unstructured systems arising from a wide range of applications.

4.1　Incomplete factorization

A variety of incomplete LU (ILU) techniques have been studied extensively in the past, including different strategies of dropping elements, such as the level-of-fill structure-based approach (i.e., ILU(0), ILU(k)) [96], the numerical threshold-based approach [95], and the numerical inverse-based multilevel approach [15].

　　A standard design strategy is to start with a complete (sparse) factorization code, modify the code to drop entries by various rules. If numerical pivoting is not a concern, the level-based dropping method can be implemented efficiently by a separate symbolic factorization phase followed by the numerical phase. The symbolic factorization phase determines the nonzero structure of the factors using the incomplete fill-path theorem [63] which is an adaptation of the fill-path theorem (Theorem 3.1) for the complete factorization.

The rationale behind the level-based method is that for some problems, the higher the level of an entry is updated, the smaller it becomes. This may not be true in general. The value-based threshold method is often more robust than the level-based method, but it is harder to implement efficiently. Here, we cannot separate the symbolic and numerical phases; they must be interleaved at each step of factorization. One of the most sophisticated value-based methods is ILUTP proposed by Saad [95, 96], which combines a dual dropping strategy with numerical pivoting ("T" stands for threshold, and "P" stands for pivoting). The dual dropping rule in ILU(τ, p) first removes the elements that are smaller than τ from the current factored row or column. It then keeps only the largest p elements to control the memory requirement.

The original ILUTP algorithm was presented as a row-wise or a column-wise variant. It suffers from the same inefficiency problem as the row-wise or column-wise LU in that it lends to very little reuse of the cached data. Recently, we developed a new variant of ILUTP that exploits *supernode* in the incomplete factors. We modified the high-performance direct solver SuperLU [29] to perform incomplete factorization. To retain supernode, a delayed dropping strategy is used which first computes the entire supernode in L, then drops some rows in the supernode if they are small in certain measure, such as vector norm [53, 71]. Although the average size of the supernodes in an incomplete factor is smaller than that in a complete factor, the supernodal ILUTP can still be twice as fast as the column-wise ILU [71]. It has the combined benefits of retaining numerical robustness of ILUTP as well as achieving fast construction and application of the ILU preconditioner.

For the secondary dropping strategy, the traditional methods examines only the current column (row), and limits the number of nonzeros allowed in this column (row). We proposed an *area-based* fill control method which examines the fill ratio due to all the preceding columns, and limits the current column size based on this dynamic fill ratio. This is shown to be more flexible and numerically robust than the column-based scheme. Furthermore, we incorporated several heuristics for adaptively modifying various threshold parameters as the factorization proceeds, which improves the robustness of the algorithm [71].

When tested with over 230 unsymmetric matrices, the supernodal, area-based adaptive ILUTP combined with GMRES can solve nearly 70% of the problems. When successful, the preconditioned iterative solver is often faster than the direct solver and uses less memory.

In general, designing an efficient ILU algorithm faces many of the similar issues as that of the complete LU algorithm. Worse yet, the ILU fatorization has even less arithmetic intensity and more sequentiality. It is extremely hard to achieve a scalable implementation. Hysom and Pothen presented a parallel approach using the domain decomposition idea [63], and showed the numerical results with 216 processors. The question remains wide open whether a parallel ILU can be designed which scales to thousands of cores, let alone millions.

4.2   Low-rank factorization

In the last 10 to 15 years, several rank structured matrix representations have been developed, such as $\mathcal{H}$-matrices [56, 59, 57], $\mathcal{H}^2$-matrices [16, 17, 58], quasiseparable matrices [13, 38], and semiseparable matrices [20, 103]. They have been widely used in the fast solutions of the integral equations and the partial differential equations using the boundary element method.

More recently we have been developing a new class of structured sparse factorization method exploiting low-rank structures using hierarchically semi-separable (HSS) matrices [19, 77, 103, 112]. The novelty of the HSS-sparse solver is to apply the HSS compression techniques to the intermediate, dense submatrices that appear in the standard sparse factorization methods, such as supernodes or frontal matrices. The resulting HSS-sparse factorization can be used as a direct solver or preconditioner depending on the application's accuracy requirement and the characteristics of the PDEs. If the randomized sampling compression technique [79] is employed in compression, it can be shown that for the 3D model problems, the HSS-sparse factorization costs $O(n)$ flops for discretized matrices from certain PDEs and $O(n^{4/3})$ for broader classes of PDEs [111]. This complexity is much lower than the $O(n^2)$ cost of the traditional, exact sparse factorization method. Moreover, the new class of HSS-sparse factorizations can be applied to much broader classes of discretized PDEs (including non-selfadjoint and indefinite ones) aiming towards optimal complexity preconditioners.

Informally, the HSS representation partitions the off-diagonal blocks of a dense matrix in a hierarchical fashion; these off-diagonal blocks are approximated by compact forms, such as truncated SVD. A key property of HSS is that the orthogonal bases are desired to be *nested* following the hierarchical partitioning. This leads to asymptotically faster construction and factorization algorithms. Figure 4.1 illustrates a block 8×8 HSS representation of A, for which the hierarchical structure and the generators U_i, V_i, R_i, and B_i are succinctly depicted by the HSS tree on the right. As a special example, its leading block 4×4 part looks like the

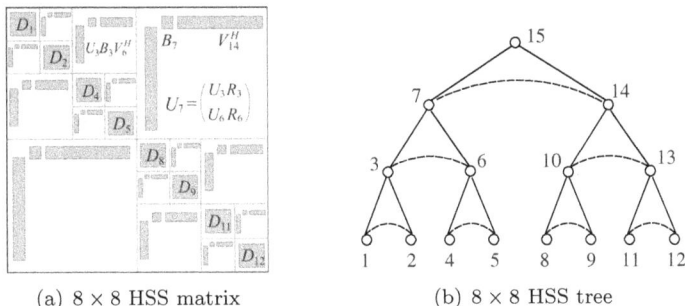

(a) 8 × 8 HSS matrix

(b) 8 × 8 HSS tree

Figure 4.1   Pictorial illustrations of a block 8 × 8 HSS form and the corresponding HSS tree $\mathcal{T}$.

following, where t_7 is the index set associated with node 7 of the HSS tree:

$$A|_{t_7 \times t_7} \approx$$

$$\left(\begin{array}{cc} \begin{pmatrix} D_1 & U_1 B_1 V_2^H \\ U_2 B_2 V_1^H & D_2 \end{pmatrix} & \begin{pmatrix} U_1 R_1 \\ U_2 R_2 \end{pmatrix} B_3 \begin{pmatrix} W_4^H V_4^H & W_5^H V_5^H \end{pmatrix} \\ \begin{pmatrix} U_4 R_4 \\ U_5 R_5 \end{pmatrix} B_6 \begin{pmatrix} W_1^H V_1^H & W_2^H V_2^H \end{pmatrix} & \begin{pmatrix} D_4 & U_4 B_4 V_5^H \\ U_5 B_5 V_4^H & D_5 \end{pmatrix} \end{array} \right).$$

With this representation, we can use the ULV factorization and the accompanying solution algorithms to solve the linear systems [21]. In [107], we developed a set of novel parallel algorithms for the key HSS operations (rank-revealing QR factorization, HSS construction, ULV factorization and HSS solution) that are used for solving dense linear systems. The parallel algorithms fully exploit both the HSS tree parallelism and dense matrix parallelism. We demonstrated that the new approach is two to 30 times faster than LU factorization; it reduces memory usage by 70- to 100-fold, using up to 8912 processing cores. Later in [106], we developed a parallel geometric HSS-embedded multifrontal sparse solver by employing the above parallel HSS algorithms to the dense frontal matrices corresponding to the separators in nested dissection. Here, we fully exploit three levels of parallelism from coarsest to finest: separator tree, HSS tree and dense matrix kernels. We tested our new parallel HSS-structured multifrontal code using up to 16,384 cores, and demonstrated that the new solver is more than 3.5 times faster than the pure multifrontal solver, and the maximum peak memory footprint is reduced by up to five-fold. Our new fast structured sparse solver has already been used successfully in several geophysics applications [108, 105]. We also show that the low-rank approximate factorization can be used as effective preconditioners for broader classes of problems [82].

5   Hybrid methods

For the three-dimensional, multiphysics, extreme scale problems, there
are a number of challenges encountered by direct solvers (e.g., large
amount of fill) and by iterative solvers (e.g., slow or no convergence). To
mitigate these difficulties, a number of parallel hybrid (direct/iterative)
solution methods have been developed [50, 41, 113, 115, 93, 33]. The
domain decomposition based hybrid methods are amenable to highly
scalable implementations. We present here a non-overlapping domain
decomposition method called the Schur complement method (a.k.a. iterative substructuring) [100]. In this method, the original linear system $Ax = b$ is first reordered, using any parallel graph partitioning
method [66, 88, 119], into a system of the following block structure:

$$
\begin{pmatrix}
D_1 & & & & E_1 \\
& D_2 & & & E_2 \\
& & \ddots & & \vdots \\
& & & D_k & E_k \\
\hline
F_1 & F_2 & \cdots & F_k & C
\end{pmatrix}
\begin{pmatrix}
u_1 \\ u_2 \\ \vdots \\ u_k \\ y
\end{pmatrix}
=
\begin{pmatrix}
f_1 \\ f_2 \\ \vdots \\ f_k \\ g
\end{pmatrix},
\tag{5.1}
$$

where D_ℓ is referred to as the ℓ-th *interior subdomain*, C consists of the
separators, and E_ℓ and F_ℓ are the *interfaces* between D_ℓ and C. The unknowns in the interior subdomains are first eliminated using techniques
from the direct solvers, and the remaining Schur complement system is
solved using a preconditioned iterative solver, such as Conjugate Gradient or GMRES, that is:

$$
Sy = \widehat{g} ,
\tag{5.2}
$$

where the Schur complement S is given by

$$
S = C - \sum_{\ell=1}^{k} F_\ell D_\ell^{-1} E_\ell ,
\tag{5.3}
$$

and $\widehat{g} = g - \sum_{\ell=1}^{k} F_\ell D_\ell^{-1} f_\ell$. The final solution vector u_ℓ is obtained by
solving the ℓ-th subdomain system

$$
D_\ell u_\ell = f_\ell - E_\ell y .
\tag{5.4}
$$

For a symmetric positive definite matrix, it can be shown that the
Schur complement has a smaller condition number than the original co-
efficient matrix [100]. Consequently, the preconditioned iterative solver
often requires fewer iterations for the Schur complement system than for
the original system. Therefore, this method has the potential of balancing the robustness of the direct solver with the efficiency of the iterative
solver since the unknowns in each interior subdomain can be eliminated

efficiently and in parallel, while the sparsity can be enforced for solving the Schur complement system, where most of fill may occur.

There are a number of challenges for developing a parallel algorithm and implementation that are both scalable and numerically robust. A straightforward parallel algorithm would assign one subdomain to each process, i.e. single-level parallelism. Then the number of subdomains must increase with the increasing number of processes, leading to a larger Schur complement S, an increase in the cost of solving (5.2) and often divergence of the iterative solver. It is thus imperative to exploit the *hierarchical parallelism*: The processes are divided into k subgroups. Each subgroup factorizes one subdomain D_ℓ in parallel using a parallel direct solver, either a shared memory one or a distributed memory one. All or a subset of processes participate in the iterative solution of the Schur complement system. Hence, the numbers of subdomains can be far fewer than the number of processes. In fact, we can keep a constant number of subdomains and the Schur complement size while increasing the number of processes. A good convergence is maintained regardless of the core count, see [113, 93] for details.

The second critical issue is to design the preconditioners for solving the Schur complement system. One method is to compute the inverses of the *local Schur complements* and form additive Schwarz preconditioner [50]. This method is more scalable but may suffer from slow convergence, especially with a large number of subdomains. The other method involves a *global approximate Schur complement*, in which some entries are dropped while forming an approximate Schur complement $\tilde{S}$. Then, a general algebraic preconditioner can be constructed using $\tilde{S}$, such as an ILU or a low-rank approximate factorization of $\tilde{S}$ (see Section 4), to precondition GMRES for solving (5.2), e.g., [41], [113], and [93]. The global method is numerically more robust, but hard to achieve a scalable implementation.

The parallel performance of the algorithm using hierarchical parallelism and global preconditioner depends on both intra-subgroup and inter-subgroup load balance. A number of new combinatorial problems arise in this context, such as multi-constraint graph partitioning and sparse matrix-matrix multiplication. Some progress was made in the area [116], but many open questions remain to be explored. Some new partitioning problems also arise in the Block Cimmino hybrid methods [31, 33].

Acknowledgments

Our special thanks go to Meiyue Shao from EPFL who provided tremendous help in setting up the parallel programming exercises on the local computer clusters and was on-site to help conduct the afternoon hands-on sessions. We thank Francois-Henry Rouet for helping to set up the VirtualBox software to host the programming exercises and assignments on everyone's laptop computers. Patrick Amestoy, Alfredo Buttari, James Demmel, John Gilbert, Jean-Yves L'Excellent, Artem Napov, Sam Williams, and Bora Uçar have graciously allowed us to use part of their lecture notes on parallel computing and sparse matrix computations.

This work was partially supported by U.S. Department of Energy, Office of Science, Advanced Scientific Computing Research under Contract No. DE-AC02-05CH11231.

References

[1] G. Amdahl. Validity of the single processor approach to achieving large-scale computing capabilities. In *AFIPS Conference Proceedings (30)*, page 483–485, Washington, D.C., April 1967.

[2] P. Amestoy, T. Davis, and I. Duff. Algorithm 837: AMD, an approximate minimum degree ordering algorithm. *ACM Trans. Mathematical Software*, 30(3): 381–388, 2004.

[3] P. R. Amestoy, T. A. Davis, and I. S. Duff. An approximate minimum degree ordering algorithm. *SIAM J. Matrix Analysis and Applications*, 17(4): 886–905, 1996. Also University of Florida TR-94-039.

[4] P. R. Amestoy and I. S. Duff. Memory management issues in sparse multifrontal methods on multiprocessors. *Int. J. Supercomputer Appl.*, 7(1): 64–82, Spring 1993.

[5] P. R. Amestoy, X. S. Li, and E. G. Ng. Diagonal markowitz scheme with local symmetrization. *SIAM Journal on Matrix Analysis and Applications*, 29(1): 228–244, 2007.

[6] P. R. Amestoy, X. S. Li, and S. Pralet. Unsymmetric ordering using a constrained markowitz scheme. *SIAM Journal on Matrix Analysis and Applications*, 29(1): 302–327, 2007.

[7] K. Asanovic, R. Bodik, B. C. Catanzaro, J. J. Gebis, P. Husbands, K. Keutzer, D. A. Patterson, W. L. Plishker, J. Shalf, S. W Williams, and K. A. Yelick. The Landscape of Parallel Computing Research: A View from Berkeley. Technical Report no. ucb/eecs-2006-183, EECS Department, Univ. of California, Berkeley, December 18 2006.

[8] C. Ashcraft. Compressed graphs and the minimum degree algorithm. *SIAM Journal on Scientific Computing*, 16: 1404–1411, 1995.

[9] Intel® Architecture Instruction Set Extensions Programming Reference. http://download-software.intel.com/sites/default/files/319433-016.pdf, 2013.

[10] S. T. Barnard and H. Simon. A fast multilevel implementation of recursive spectral bisection for partitioning unstructured problems. In R. F. Sincovec, D. Keyes, M. Leuze, L. Petzold, and D. Reed, editors, *Sixth*

SIAM conference on parallel processing for scientific computing, pages 711–718, 1993.

[11] R. Barrett, M. Berry, T. F. Chan, J. Demmel, J. Donato, J. Dongarra, V. Eijkhout, R. Pozo, C. Romine, and H. van der Vorst. *Templates for the solution of linear systems: Building blocks for the iterative methods, 2nd Edition*. SIAM, Philadelphia, PA, 1994.

[12] Berkeley Benchmark and Optimization group. http://bebop.cs.berkeley.edu.

[13] T. Bella, Y. Eidelman, and V. Gohberg, I. and. Olshevsky. Computations with quasiseparable polynomials and matrices. *Theoret. Comput. Sci.*, 409: 158–179, 2008.

[14] M. Benzi. Preconditioning techniques for large linear systems: A survey. *J. Comp. Phys.*, 182: 418–477, 2002.

[15] M. Bollhöfer and Y. Saad. Multilevel preconditioners constructed from inverse-based ILUs. *SIAM J. Scientific Computing*, 27(5): 1627–1650, 2006.

[16] S. Börm, L. Grasedyck, and W. Hackbusch. Introduction to hierarchical matrices with applications. *Eng. Anal. Bound. Elem*, 27: 405–422, 2003.

[17] S. Börm and W. Hackbusch. Data-sparse approximation by adaptive $\mathcal{H}^2$-matrices. *Technical report, Leipzig, Germany: Max Planck Institute for Mathematics*, 86, 2001.

[18] T. Chan, J. Gilbert, and S.-H. Teng. Geometric spectral partitioning. Technical Report CSL-94-15, Palo Alto Research Center, Xerox Corporation, California, 1994.

[19] S. Chandrasekaran, P. Dewilde, M. Gu, W. Lyons, and T. Pals. A fast solver for hss representations via sparse matrices. *SIAM J. Matrix Anal. Appl.*, 29: 67–81, 2006.

[20] S. Chandrasekaran, P. Dewilde, M. Gu, T. Pals, X. Sun, A.-J. van der Veen, and D. White. Some fast algorithms for sequentially semiseparable representations. *SIAM J. Matrix Anal. Appl.*, 27: 341–364, 2005.

[21] S. Chandrasekaran, M. Gu, and T. Pals. A fast *ULV* decomposition solver for hierarchically semiseparable representations. *SIAM J. Matrix Anal. Appl.*, 28: 603–622, 2006.

[22] T.-Y. Chen, J. Gilbert, and S. Toledo. Toward an efficient column minimum degree code for symmetric multiprocessors. In *Proceedings of the 9th SIAM Conference on Parallel Processing for Scientific Computing*, San Antonio, Texas, 1999.

[23] C. Chevalier and F. Pellegrini. PT-Scotch: A tool for efficient parallel graph ordering. *Parallel Computing*, 34(6-8): 318–331, 2008.

[24] P. Colella. Defining software requirements for scientific computing, presentation.

[25] CUDA Parallel Computing Platform. http://www.nvidia.com/object/cuda_home_new.html.

[26] E. Cuthill and J. McKee. Reducing the bandwidth of sparse symmetric matrices. In *Proc. ACM Nat. Conf.*, pages 157–172, New York, 1969.

132   X. S. Li

[27] T. A. Davis. Algorithm 915, SuiteSparseQR: Multifrontal multithreaded rank-revealing sparse QR factorization. *ACM Trans. Mathematical Software*, 38(1), 2011. http://www.cise.ufl.edu/research/sparse/SPQR/.

[28] T. A. Davis, J. R. Gilbert, S. Larimore, and E. Ng. A column approximate minimum degree ordering algorithm. *ACM Trans. Mathematical Software*, 30(3): 353–376, 2004.

[29] J. W. Demmel, S. C. Eisenstat, J. R. Gilbert, X. S. Li, and J. W. H. Liu. A supernodal approach to sparse partial pivoting. *SIAM J. Matrix Analysis and Applications*, 20(3): 720–755, 1999.

[30] J. W. Demmel, J. R. Gilbert, and X. S. Li. An asynchronous parallel supernodal algorithm for sparse gaussian elimination. *SIAM J. Matrix Analysis and Applications*, 20(4): 915–952, 1999.

[31] L. A. Drummond, I. S. Duff, R. Guivarch, D. Ruiz, and M. Zenadi. Partitioning strategies for the block cimmino algorithm. Technical Report RAL-P-2013-010, RAL, 2013.

[32] I. S. Duff. MA28 - A set of FORTRAN subroutines for sparse unsymmetric linear equations. Technical Report AERE R-8730, Harwell, 1977.

[33] I. S. Duff, R. Guivarch, D. Ruiz, and M. Zenadi. The augmented block cimmino distributed method. Technical Report RAL-P-2013-001, RAL, 2013.

[34] I. S. Duff and J. K. Reid. MA47, a Fortran code for direct solution of indefinite sparse symmetric linear systems. Technical Report RAL-95-001, Rutherford Appleton Laboratory, 1995.

[35] I. S. Duff and J. K. Reid. The design of MA48, a code for the direct solution of sparse unsymmetric linear systems of equations. *ACM Trans. Mathematical Software*, 22: 187–226, 1996.

[36] I. S. Duff and J. K. Reid. The multifrontal solution of indefinite sparse symmetric linear equations. *ACM Trans. Mathematical Software*, 9(3): 302–325, September 1983.

[37] A. L. Dulmage and N. S. Mendelsohn. Two algorithms for bipartite graphs. *J. Sot. Indust. Appl. Math.*, 11: 183–194, 1963.

[38] Y. Eidelman and I. Gohberg. On a new class of structured matrices. *Integral Equations Operator Theory*, 34: 293–324, 1999.

[39] S. C. Eisenstat and J. W. H. Liu. Exploiting structural symmetry in sparse unsymmetric symbolic factorization. *SIAM J. Matrix Anal. Appl.*, pages 13: 202–211, 1992.

[40] S. C. Eisenstat, M. H. Schultz, and A. H. Sherman. Applications of an element model for gaussian elimination. In J.R. Bunch and D.J. Rose, editors, *Sparse Matrix Computations*, pages 85–96. Academic Press, 1976.

[41] J. Gaidamour and P. Henon. HIPS: a parallel hybrid direct/iterative solver based on a schur complement. In *Proc. PMAA*, 2008.

[42] A. George. Nested dissection of a regular finite element mesh. *SIAM J. Numerical Analysis*, 10: 345–363, 1973.

[43] A. George, M. T. Heath, E. Ng, and J. W. H. Liu. Symbolic cholesky factorization on a local-memory multiprocessor. *Parallel Comput.*, 5: 85–95, 1987.

[44] A. George and J. W. H. Liu. *Computer Solution of Large Sparse Positive Definite Systems.* Prentice Hall, Englewood Cliffs, NJ, 1981.

[45] A. George and J. W. H. Liu. The evolution of the minimum degree ordering algorithms. *SIAM Review*, 31(1): 1–19, March 1989.

[46] A. George and E. Ng. An implementation of Gaussian elimination with partial pivoting for sparse systems. *SIAM J. Sci. Stat. Comput.*, 6(2): 390–409, 1985.

[47] J. A. George and J. W. H. Liu. A fast implementation of the minimum degree algorithm using quotient graphs. *ACM Trans. Mathematical Software*, 6: 337–358, 1980.

[48] J. R. Gilbert and R. E. Tarjan. The analysis of a nested dissection algorithm. *Numerische Mathematik*, pages 377–404, 1987.

[49] John R. Gilbert and Esmond G. Ng. Predicting structure in nonsymmetric sparse matrix factorizations. In A. George, J. R. Gilbert, and J. W. H. Liu, editors, *Graph theory and sparse matrix computation*, pages 107–139. Springer-Verlag, New York, 1993.

[50] L. Giraud, A. Haidar, and S. Pralet. Using multiple levels of parallelism to enhance the performance of domain decomposition solvers. *Parallel Computing*, 36: 285–296, 2010.

[51] L. Grigori, E. Boman, S. Donfack, and T. Davis. Hypergraph-based unsymmetric nested dissection ordering for sparse lu factorization. *SIAM J. Scientific Computing*, 32(6), 2010.

[52] L. Grigori, J. W. Demmel, and X. S. Li. Parallel symbolic factorization for sparse LU with static pivoting. *SIAM J. Scientific Computing*, 29(3): 1289–1314, 2007.

[53] A. Gupta and T. George. Adaptive techniques for improving the performance of incomplete factorization preconditioning. *SIAM J. Sci. Comput.*, 32(1): 84–110, 2010.

[54] A. Gupta, F. Gustavson, M. Joshi, G. Karypis, and V. Kumar. Design and implementation of a scalable parallel direct solver for sparse symmetric positive definite systems. In *Proceedings of the 8th SIAM Conference on Parallel Processing for Scientific Computing*, Minneapolis, Minnesota, 1997.

[55] A. Gupta and V. Kumar. Optimally scalable parallel sparse cholesky factorization. In *The 7th SIAM Conference on Parallel Processing for Scientific Computing*, pages 442–447, 1995.

[56] W. Hackbusch. A sparse matrix arithmetic based on $\mathcal{H}$-matrices. Part I: introduction to $\mathcal{H}$-matrices. *Computing*, 62: 89–108, 1999.

[57] W. Hackbusch, L. Grasedyck, and S. Börm. An introduction to hierarchical matrices. *Math. Bohem.*, 127: 229–241, 2002.

[58] W. Hackbusch, B. Khoromskij, and S. A. Sauter. On $\mathcal{H}^2$-matrices. *Lectures on applied mathematics (Munich, 1999), Springer, Berlin*, pages 9–29, 2000.

[59] W. Hackbusch and B. N. Khoromskij. A sparse $\mathcal{H}$-matrix arithmetic. Part-II: Application to multi-dimensional problems. *Computing*, 64: 21–47, 2000.

[60] B. Hendrickson and R. Leland. A multilevel algorithm for partitioning graphs. In *Proc. Supercomputing*, San Diego, CA, 1995.

[61] B. Hendrickson and R. Leland. The CHACO's User's Guide. Technical Report SAND95-2344•UC-405, Sandia National Laboratories, Albuquerque, 1995. http://www.cs.sandia.gov/~bahendr/chaco.html.

[62] J. L. Hennessy and D. A. Patterson. *Computer Architecture: A Quantitative Approach, Fifth Edition.* Elsevier, 2012.

[63] D. Hysom and A. Pothen. A scalable parallel algorithm for incomplete factor preconditioning. *SIAM J. Scientific Computing*, 22(6): 2194–2215, 2001.

[64] G. Karypis and V. Kumar. MEIIS – serial graph partitioning and computing fill-reducing matrix ordering. University of Minnesota. http://glaros.dtc.umn.edu/gkhome/views/metis.

[65] G. Karypis and V. Kumar. A fast and high quality multilevel scheme for partitioning irregular graphs. *SIAM J. Scientific Computing*, 20: 359–392, 1998.

[66] G. Karypis, K. Schloegel, and V. Kumar. PARMEIIS: Parallel graph partitioning and sparse matrix ordering library – version 3.1. University of Minnesota, 2003. http://www-users.cs.umn.edu/~karypis/metis/parmetis/.

[67] G. Krawezik and G. Poole. Accelerating the ANSYS direct sparse solver with GPUs. In *Proc. Symposium on Application Accelerators in High Performance Computing (SAAHPC)*, Urbana-Champaign, IL, NCSA, 2009. http://saahpc.ncsa.illinois.edu/09.

[68] P. S. Kumar, M. K. Kumar, and A. Basu. A parallel algorithm for elimination tree computation and symbolic factorization. *Parallel Comput.*, 18: 849–856, 1992.

[69] X. S. Li. Evaluation of sparse factorization and triangular solution on multicore architectures. In *Proceedings of VECPAR'08 8th International Meeting High Performance Computing for Computational Science*, Toulouse, France, June 24-27 2008.

[70] X. S. Li and J. W. Demmel. SuperLU_DIST: A scalable distributed-memory sparse direct solver for unsymmetric linear systems. *ACM Trans. Mathematical Software*, 29(2): 110–140, June 2003.

[71] X. S. Li and M. Shao. A supernodal approach to imcomplete LU factorization with partial pivoting. *ACM Trans. Mathematical Software*, 37(4), 2010.

[72] R. J. Lipton, D. J. Rose, and R. E. Tarjan. Generalized nested dissection. *SIAM J. Numerical Analysis*, 16: 346–358, 1979.

[73] J. W. H. Liu. Modification of the minimum degree algorithm by multiple elimination. *ACM Trans. Mathematical Software*, 11: 141–153, 1985.

[74] J. W. H. Liu. The role of elimination trees in sparse factorization. *SIAM J. Matrix Analysis and Applications*, 11: 134–172, 1990.

[75] J. W. H. Liu. The multifrontal method for sparse matrix solution: theory and practice. *SIAM Review*, 34(1): 82–109, March 1992.

[76] R. Lucas, G. Wagenbreth, D. Davis, and R. Grimes. Multifrontal computations on GPUs and their multi-core hosts. In *VECPAR'10: Proc. 9th Intl. Meeting for High Performance Computing for Computational Science*, Berkeley, CA, 2010. http://vecpar.fe.up.pt/2010/papers/5.php.

[77] W. Lyons. *Fast Algorithms with Applications to PDEs*. PhD thesis, University of California, Santa Barbara, 2005.

[78] H. M. Markowitz. The elimination form of the inverse and its application to linear programming. *Management Sci.*, 3: 255–269, 1957.

[79] P. G. Martinsson. A fast randomized algorithm for computing a hierarchically semiseparable representation of a matrix. *SIAM J. Matrix Analysis and Applications*, 32(4): 1251–1274, 2011.

[80] Message Passing Interface (MPI) forum. http://www.mpi-forum.org/.

[81] Mpi tutorial. https://computing.llnl.gov/tutorials/mpi/.

[82] A. Napov, X. S. Li, and M. Gu. An algebraic multifrontal preconditioner that exploits the low-rank property. Technical report, Lawrence Berkeley National Laboratory, 2014. In preparation.

[83] L. Oliker, X. Li, P. Husbands, and R. Biswas. Effects of ordering strategies and programming paradigms on sparse matrix computations. *Siam Review*, 44(3): 373–393, 2002.

[84] Open Computing Language. http://www.khronos.org/opencl/.

[85] OpenMP API Specification for Parallel Programming. www.openmp.org.

[86] OpenMP Tutorial. https://computing.llnl.gov/tutorials/openMP/.

[87] G. Pagallo and C. Maulino. A bipartite quotient graph model for unsymmetric matrices. In *Lecture Notes in Mathematics 1005, Numerical Method*, pages 227–239, Springer-Verlag, New York, 1983.

[88] F. Pellegrini. PT-Scotch and libScotch 5.1 User's Guide (version 5.1.11). INRIA Bordeaux Sud-Ouest, Université Bordeaux I. November, 2010. http://www.labri.fr/perso/pelegrin/scotch/.

[89] F. Pellegrini. Scotch and libScotch 5.1 User's Guide (version 5.1.11). INRIA Bordeaux Sud-Ouest, Université Bordeaux I. November, 2010. http://www.labri.fr/perso/pelegrin/scotch/.

[90] F. Pellegrini, J. Roman, and P. Amestoy. Hybridizing Nested Dissection and Halo Approximate Minimum Degree for Efficient Sparse Matrix Ordering. *Concurrency: Practice and Experience*, 12: 69–84, 2000.

[91] A. Pothen and C.-J. Fan. Computing the block triangular form of a sparse matrix. *ACM Trans. Mathematical Software*, Vol. 16: 303–324, 1990.

[92] A. Pothen, H. D. Simon, L. Wang, and S. Barnard. Towards a fast implementation of spectral nested dissection. In *Supercomputing, ACM Press*, pages 42–51, 1992.

[93] S. Rajamanickam, E. G. Boman, and M. A. Heroux. ShyLU: A hybrid-hybrid solver for multicore platforms. In *Proc. of IPDPS 2012*, Shanghai, China, May 20-24 2012.

[94] D. J. Rose and R. E. Tarjan. Algorithmic aspects of vertex elimination of directed graphs. *SIAM Journal on Applied Math*, Vol. 34(No. 1): 176–197, January 1978.

[95] Y. Saad. ILUT: A dual threshold incomplete LU factorization. *Numerical Linear Algebra with Applications*, 1(4): 387–402, 1994.

[96] Y. Saad. *Iterative methods for sparse linear systems*. SIAM, Philadelphia, 2004.

[97] P. Sao, R. Vuduc, and X. Li. A distributed CPU-GPU sparse direct solver. In *Euro-Par 2014*, Porto, Portugal, August 25–29, 2014.

[98] O. Schenk, K. Gärtner, and W. Fichtner. Efficient sparse LU factorization with left–right looking strategy on shared memory multiprocessors. *BIT*, 40(1): 158–176, 2000.

[99] R. Schreiber. A new implementation of sparse Gaussian elimination. *ACM Trans. Mathematical Software*, 8: 256–276, 1982.

[100] B. Smith, P. Bjorstad, and W. Gropp. *Domain Decomposition. Parallel Multilevel Methods for Elliptic Partial Differential Equations*. Cambridge University Press, New York, 1996.

[101] W. F. Tinney and J. W. Walker. Direct solutions of sparse network equations by optimally ordered triangular factorization. *Proc. IEEE*, 55: 1801–1809, 1967.

[102] B. Uçar and C. Aykanat. Revisiting hypergraph models for sparse matrix partitioning. *SIAM Review*, 49(4): 595–603, 2007.

[103] R. Vandebril, M. Van Barel, G. Golub, and N. Mastronardi. A bibliography on semiseparable matrices. *Calcolo*, 42: 249–270, 2005.

[104] B. Vastenhouw and R. H. Bisseling. A two-dimensional data distribution method for parallel sparse matrix-vector multiplication. *SIAM Review*, 47(1): 67–95, 2005.

[105] S. Wang, M. V. de Hoop, J. Xia, and X. S. Li. Massively parallel structured multifrontal solver for time-harmonic elastic waves in 3D anisotropic media. *Geophysics J. Int.*, 191: 346–366, 2012.

[106] S. Wang, X. S. Li, F.-H. Rouet, J. Xia, and M. V. de Hoop. A parallel fast geometric multifrontal solver using hierarchically semiseparable structure. *ACM Trans. Mathematical Software*, 2013. (submitted).

[107] S. Wang, X. S. Li, J. Xia, Y. Situ, and M. V. de Hoop. Efficient parallel algorithms for solving linear systems with hierarchically semiseparable structures. *SIAM J. Scientific Computing*, 35(6): C519–C544, 2013.

[108] S. Wang, J. Xia, M. V. de Hoop, and X. S. Li. Massively parallel structured direct solver for equations describing time-harmonic qp-polarized waves in tti media. *Geophysics*, 77: T69–T82, 2012.

[109] S. Williams, A. Waterman, and D. Patterson. Roofline: An Insightful Visual Performance Model for Floating-Point Programs and Multicore Architectures. *Communications of the ACM (CACM)*, April 2009.

[110] M. M. Wolf, E. G. Boman, and B. Hendrickson. Optimizing parallel sparse matrix-vector multiplication by corner partitioning. In *PARA08*, Trondheim, Norway, May 2008.

[111] J. Xia. Randomized sparse direct solvers. *SIAM J. Matrix Anal. Appl.*, 34(1): 197–227, 2013.

[112] J. Xia, S. Chandrasekaran, M. Gu, and X. S. Li. Fast algorithms for hierarchically semiseparable matrices. *Numer. Linear Algebra Appl.*, 2010: 953–976, 2010.

[113] I. Yamazaki and X. S. Li. On techniques to improve robustness and scalability of the schur complement method. In *Proc. of VECPAR 2010*, Berkeley, California, 2010.

[114] I. Yamazaki and X. S. Li. New scheduling strategies and hybrid programming for a parallel right-looking sparse LU factorization on multicore cluster systems. In *Proceedings of IEEE International Parallel and Distributed Processing Symposium (IPDPS 2012)*, Shanghai, China, May 20-24 2012.

[115] I. Yamazaki, X. S. Li, F.-H. Rouet, and B. Uçar. Partitioning, ordering, and load balancing in a hierarchically parallel hybrid linear solver. *Int'l J. of High Performance Comput.*, 2012 (revised).

[116] I. Yamazaki, X. S. Li, F.-H. Rouet, and B. Uçar. On partitioning and reordering problems in a hierarchically parallel hybrid linear solver. In *Proc. of IPDPS 2013 Workshops*, Boston, May 20-24 2013. PDSEC-13 Workshop.

[117] M. Yannakakis. Computing the minimum fill-in is NP-complete. *SIAM J . Alg. & Disc. Meth.*, 2: 77–79, 1981.

[118] C. D. Yu, W. Wang, and D. Pierce. A CPU-GPU hybrid approach for the unsymmetric multifrontal method. *Parallel Computing*, 37: 759–770, 2011.

[119] Zoltan: Parallel partitioning, load balancing and data-management services. http://www.cs.sandia.gov/zoltan/Zoltan.html.

www.ingramcontent.com/pod-product-compliance
Lightning Source LLC
Chambersburg PA
CBHW050643190326
41458CB00008B/2398